WITHOUT EXCUSE

SEEING GOD THROUGH THE LENS OF SCIENCE

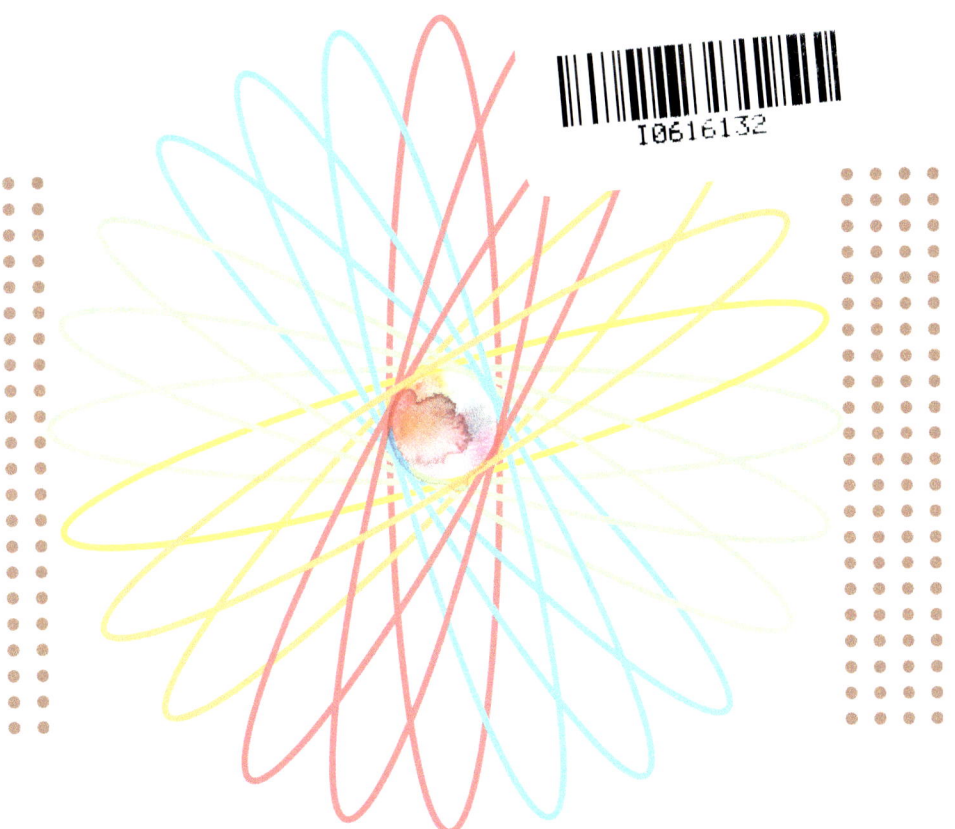

I0616132

For since the creation of the world God's invisible qualities, his eternal power and divine nature, have been clearly seen, being understood from what has been made, so that men are **without excuse.** *Romans 1:20*

BY TRISTA MCREYNOLDS
Edited by Geoffrey McReynolds and Gentry Quehl

WITHOUT EXCUSE

First RELEASE August 2025

All Scripture quotations, unless otherwise indicated, are taken from the Holy Bible: New International Version.
No AI was used in the writing of this text.

Loyston Point Press
Andersonville, Tennessee

Books by Trista McReynolds
_____

small dog BIG UNIVERSE
One Little Football Player

Available where books are sold
eBook versions on Kindle and Apple

Dedicated to all my former and future students
and
In memory of the ones who went on ahead
Marcus Ramsey King
Austin James McReynolds
Tanner Douglas Hill
Jaxon Douglas Best
See you soon enough!

Wrong will be right, when Aslan comes in sight,
At the sound of his roar; sorrow will be no more.
When he bares his teeth, winter meets its death,
And when he shakes his mane, we shall have spring again.
Mr. Beaver, in The Lion, the Witch and the Wardrobe by C.S. Lewis

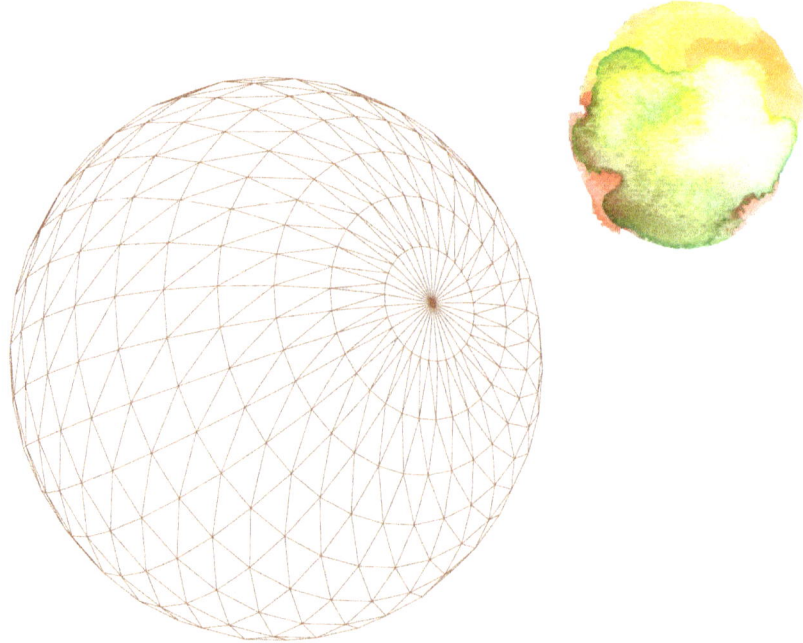

Table of Contents

Preface

Physical science is a branch of science that tries to explain how and why energy and matter function in our world. This book attempts to explore some of the concepts found in the fields of chemistry and physics; to explain them simply and reveal the Creator through the lens of His creation.

My prayer is that by the end of this journey, you will have seen clearly that this orderly world has indeed been made by a benevolent, loving God who wants nothing more than to have a relationship with you, His most prized creation.

Part 1
Chemistry

Chemistry is the study of what things are made of, how those things are put together, what their characteristics are, and how things interact with each other.

1 Matter

We think there is color, we think there is sweet, we think there is bitter, but in reality, there are atoms and a void. - Democritus, Greek philosopher

Matter is anything that has mass and takes up space. Everything in our physical world is considered matter except for a few things like light, heat, and sound. Matter can be measured in various ways. We use scientific tools to find the mass, weight, volume, or density of matter.

We are made of matter. We take up space and have mass. As human beings, made in the image of God, we are more than just a physical body, composed of flesh and bone that can be measured. The bible teaches that God is a trinity, and being made in His image makes us a trinity as well. We have a body, a soul, and a spirit.

Our soul and spirit desire a relationship with God. When we believe in Him, putting our faith and hope in Him, our spirit man begins to grow. As our spirit grows, so grows our soul. We gain knowledge of right and wrong. We become wise, learning to control our thoughts and feelings, and then we become better at managing our bodies.

The apostle Peter reminds us of the importance of growing up in "increasing measure" (2 Peter 1:8) while we live in this 'tent of a body' (2 Peter 1:13) — as our body grows, we must also be about developing our soul and spirit. A great reminder to make what truly matters matter.

Ephesians 4:14-15
Then we will no longer be infants, tossed back and forth by the waves, and blown here and there by every wind of teaching and by cunning and craftiness of men in their deceitful scheming. Instead, speaking the truth in love, we will in all things grow up into him who is the Head, that is, Christ.

Ephesians 5:15-17
Be very careful, then, how you live–not as unwise but as wise, making the most of every opportunity, because the days are evil. Therefore, do not be foolish, but understand what the Lord's will is.

Colossians 1:6-7
All over the world, the gospel is bearing fruit and growing, just as it has been doing among you since the day you heard it and understood God's grace in all its truth.

1 Peter 2:2-3
Like newborn babies, crave pure spiritual milk, so that by it you may grow up in your salvation, now that you have tasted that the Lord is good.

2 Peter 1:3-4
His divine power has given us everything we need for life and godliness through our knowledge of him who called us by his own glory and goodness. Through these he has given us his very great and precious promises, so that through them you may participate in the divine nature and escape the corruption in the world caused by evil desires.

2 The Atom

Matter, though divisible in an extreme degree, is nevertheless not infinitely divisible. That is, there must be some point beyond which we cannot go in the division of matter...I have chosen the word "atom" to signify these ultimate particles.
- John Dalton, English chemist and physicist

The basic building block of all matter is the atom. 118 unique atoms make up everything in our physical world. Atoms are so small that it is said one million atoms can be lined up side by side and equal the width of a single human hair.

Stars, water, and salt are types of matter. Stars are made of the atoms hydrogen and helium. Water is made of the atoms of hydrogen and oxygen. Salt is a combination of sodium and chlorine atoms. All of the 118 known atoms combine in so many infinite ways that we have everything we need to live.

In Genesis chapters 1 and 2, we find the account of creation. Each day of creation brought about new arrangements and combinations of atoms to create nonliving and living things. God created everything in the universe: from the atmosphere, land, and seas, to the seed-bearing plants and trees, the sun, moon, and stars; all kinds of animals; saving His best for last -- humans.

While Earth was growing new life, God set the masterpiece of His creation, human beings, in a beautiful garden. Adam and Eve worked and tended the garden. God walked with them and they enjoyed fellowship and plenty.

Like Adam and Eve, God made us and has a purpose for each one of us. May we fellowship with our Creator and build our lives upon Him.

Genesis 1:27
So God created man in his own image, in the image of God he created him; male and female he created them.

Genesis 5:1-2
When God created man, he made him in the likeness of God. He created them male and female and blessed them. And when they were created, he called them man.

Psalms 89:11
The heavens are yours, and yours also the earth; you founded the world and all that is in it.

Isaiah 42:5-6a
This is what God the Lord says – he who created the heavens and stretched them out, who spread out the earth and all that comes out of it, who gives breath to its people, and life to those who walk on it: "I, the Lord, have called you in righteousness; I will take hold of your hand."

Colossians 1:15-17
He (Jesus) is the image of the invisible God, the firstborn over all creation. For by him all things were created: things in heaven and on earth, visible and invisible, whether thrones or powers or rulers or authorities; all things were created by him and for him. He is before all things and in him all things hold together.

3 Protons

An atom is identified by the number of protons it contains. Protons are the positively charged subatomic particles found in the nucleus of all atoms. If an atom has 29 protons, it is identified or known as Copper. If an atom has 79 protons, it is Gold. If an atom has one less proton or one more proton it has a different identity.

When an atom's proton count changes, the very essence of the atom changes. The atom no longer looks, feels, and responds to other atoms the same way because it is a different atom.

In Genesis Chapter 3, we read the account of the fall of man. Sin entered the perfect world God created, and man lost his true identity. Humans forgot their purpose and lost the authority and dominion God had given them. They lost their essence or identity.

But, in God's infinite wisdom, amazing grace, and never-ending love, His only Son, Jesus, agreed to come to earth in the form of a man to restore what had been lost.

When we recognize the sinful nature within us, repent of it, and believe in the name of Jesus, our identity changes. We become who we were created to be - a son or daughter of the Most High, created in the image of God, full of the characteristics and behaviors of Him.

Romans 3:23
For all have sinned and fall short of the glory of God.

Romans 5:8
But God demonstrates his own love for us in this: While we were still sinners, Christ died for us.

Romans 6:23
For the wages of sin is death, but the gift of God is eternal life in Christ Jesus our Lord.

Romans 10:9
That if you confess with your mouth, "Jesus is Lord," and believe in your heart that God raised him from the dead, you will be saved.

2 Corinthians 5:17-18
Therefore, if anyone is in Christ, he is a new creation; the old has gone, the new has come! All this is from God, who reconciled us to himself through Christ and gave us the ministry of reconciliation.

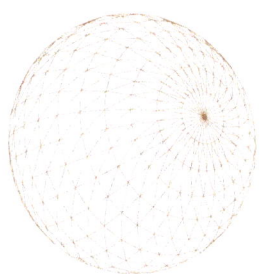

4 Neutrons

After the knowledge of and obedience to the will of God, the next aim must be to know something of His attributes of wisdom, power, and goodness as evidenced by his handiwork. - James Prescott Joule, English physicist

Neutrons are the largest of the three parts of an atom. They have a neutral charge and add mass to the atom. The main job of the neutron is to provide stability to the atom. Neutrons are found in the center of each atom along with the protons. The combination of neutrons and protons is known as the nucleus. Because all protons have a positive charge, they tend to repel each other, but the neutral neutrons mixed in with the protons lessen the repulsion of the protons.

We tend to push people away; to repel one another like the proton. Because humans have the sin nature to overcome, God has provided a helper; someone to provide stability to our lives: The Holy Spirit. Jesus told his disciples that the Father would send the Holy Spirit to us when Jesus ascended back to heaven, and He said it would be better for us to have Him with us.

The Holy Spirit is within us to help us remember the teachings of Jesus. He convicts us of sin, reminds us of our salvation, and gives us boldness to share the gospel with others. The Holy Spirit is our confidence, which allows us to live at peace with those who would normally repel us.

Luke 24:49
I am going to send you what my Father has promised, but stay in the city until you have been clothed with power from on high.

John 14:26-27
But the Counselor, the Holy Spirit, whom the Father will send in my name, will teach you all things and will remind you of everything I have said to you. Peace I leave you; my peace I give you. I do not give to you as the world gives. Do not let your hearts be troubled and do not be afraid.

John 16:7-8
But I tell you the truth: it is for your good that I am going away. Unless I go away, the counselor will not come to you, but if I go, I will send him to you. When he comes, he will convict the world of guilt in regard to sin and righteousness and judgment.

John 16:13-14
But when he, the Spirit of truth, comes, he will guide you into all truth. He will not speak on his own; he will speak only what he hears, and he will tell you what is yet to come. He will bring glory to me by taking from is mine and making it known to you.

Acts 2:4-5
On one occasion, while he was eating with them, he gave them this command: "Do not leave Jerusalem, but wait for the gift my father promised, which you have heard me speak about. For John baptized with water, but in a few days you will be baptized with the Holy Spirit."

5 Isotopes

Question: What did the scientist say when he found 2 isotopes of helium?
Answer: He He

Isotopes are atoms with the same identity; they have the same number of protons, but different amounts of neutrons. Since the neutron is the largest subatomic particle, it has the most mass. When the total mass of an atom changes, it causes the atom to express its physical properties differently. For example, a neutral carbon atom is considered stable, but when two more neutrons are present, the carbon atom becomes unstable and radioactive.

A similar change happens to us when we get weighted down with a load we were never meant to carry. We become heavy with burdens, worry, fear, and weariness. Our outward countenance and attitude change; we become unstable, angry, short-tempered, and hurtful to those around us.

The book of Proverbs tells us that an anxious heart weighs a man down, and Jesus tells us to be careful, or our hearts will be weighed down with the anxieties of life. But there is good news. We don't have to carry the weight.

In Matthew 11:28-30, Jesus says, "Come to me, all you who are weary and burdened, and I will give you rest. Take my yoke upon you and learn from me, for I am gentle and humble in heart, and you will find rest for your souls. For my yoke is easy and my burden is light.'

Psalms 55:22
Cast your cares upon the Lord, and he will sustain you; he will never let the righteous fall.

Proverbs 12:25
An anxious heart weighs a man down, but a kind word cheers him up.

Matthew 6:27
Who of you by worrying can add a single hour to his life?

Matthew 6:33-34
But seek first his kingdom and his righteousness, and all these things will be given to you as well. Therefore do not worry about tomorrow, for tomorrow will worry about itself. Each day has enough trouble of its own.

1 Peter 5:7
Cast all your anxiety on him because he cares for you.

6 Electrons

Everything we call real is made of the things that cannot be regarded as real.
- Niels Bohr, Danish physicist

The negatively charged electron is the part of the atom that orbits the nucleus, located in the center of the atom. These tiny subatomic particles are stable and active. The number of electrons in an atom is equal to the number of its protons. The arrangement of these electrons influences the atom's chemical properties. The outermost electrons determine how the atom bonds or reacts with other atoms.

Our outward appearance gives the first impression of who we may be and how people react to us. When we spend time with the Lord, there is a glow about us that others notice.

There is an account of Moses in the book of Exodus. When Moses came down from Mount Sinai with the two tablets of the commands God had given, he was not aware that his face was radiant because he had spoken to the Lord. His face was so bright that the people were afraid of him.

The bible talks about the glory of God, and the more time we spend with God, the more of His glory rubs off on us. We start looking different, and others notice. To some people we meet, this 'glory glow' may be something that causes them to look at us funny or be a bit distant, but I have found that a smile and a simple hello usually breaks the ice and allows us to share a bit of His glory with them.

Exodus 34:29

When Moses came down from Mount Sinai with two tablets of the testimony in his hands, he was not aware that his face was radiant because he had spoken with the Lord.

Psalms 31:16

Let your face shine on your servant; save me in your unfailing love.

Isaiah 60:1-2

Arise, shine, for your light has come, and the glory of the Lord rises upon you. See, darkness covers the earth and thick darkness is over the peoples, but the Lord rises upon you.

Matthew 5:14-16

You are the light of the world. A city on a hill cannot be hidden. Neither do people light a lamp and put it under a bowl. Instead they put it on a stand, and it gives light to everyone in the house. In the same way, let your light shine before men, that they may see your good deeds and praise your Father in heaven.

Acts 6:15

All who were sitting in the Sanhedrin looked intently at Stephen, and they saw that his face was like the face of an angel.

7 Atomic Radius and Electronegativity

"...I found it difficult to imagine that there could be a real conflict between scientific truth and spiritual truth. Truth is truth. Truth cannot disprove truth."
- Francis Collins, American physician-geneticist

Scientists have discovered that atoms have specific sizes. These tiny spheres can be measured by finding the distance from the nucleus to the outermost energy level. This measurement is called the atomic radius. The atomic radius is linked to an atom's electronegativity.

The atomic electronegativity is a force found within the nucleus of an atom that attracts and pulls the electrons towards the nucleus. Scientists have found that the smaller the atomic radius, the stronger the electronegative force within the atom. In other words, the smaller the atom is, the 'stronger' it is.

In the bible, we see a similar truth; Philippians 2:6-10 says that Jesus humbled Himself - made Himself nothing - and God exalted Him. We are told that our attitude should be the same as Christ's. When we make ourselves 'small' through the act of humility, we submit our attitudes and actions to the ways of God.

It's a hard thing to overcome pride and let go, because it feels like we will be taken advantage of or seem weak, but God promises that He will lift us up and make us strong if we put aside our pride and embrace humility.

2 Corinthians 12:10
That is why, for Christ's sake, I delight in weakness, in insults, in hardships, in persecutions, in difficulties. For when I am weak, then I am strong.

Ephesians 6:10
Finally, be strong in the Lord and in his mighty power.

Philippians 4:13
I can do everything through him who gives me strength.

James 4:10
Humble yourselves before the Lord, and he will lift you up.

1 Peter 5:6
Humble yourselves, therefore, under God's mighty hand, that he may lift you up in due time.

8 Strong Nuclear Force

When we have found how the nucleus of atoms is built up, we shall have found the greatest secret of all - except life. - Ernest Rutherford, New Zealand physicist

Have you ever wondered how protons and neutrons stay together to make the nucleus of the atom? Scientists believe there is a type of pulling force between the two particles known as the strong nuclear force. It is believed to be one of the four strongest forces in our universe.

An atomic bomb is an example of the potential power of the strong nuclear force held within the nucleus of atoms. Part of the energy released from an atomic bomb comes from separating the protons and neutrons, a process referred to as nuclear fission. When the nucleus splits, it releases the neutrons and creates a chain reaction, leading to a huge amount of energy that is released. This powerful energy holding the nucleus together is nothing compared to the powerful promises of God.

In 2 Corinthians 1:20-22, we find that no matter how many promises God has made, they are "Yes" in Christ. God's yes and our yes together is a force so strong that it binds us to Him from the day we say, "Yes," throughout all eternity.

Take heart today and think on this: We have God's eternal pledge. He has anointed us, set His seal of ownership on us, and put H Spirit in our hearts as a deposit guaranteeing what is to come. Amen.

Romans 8:35
Who shall separate us from the love of Christ? Shall trouble or hardship or persecution or famine or nakedness or danger or sword?

2 Corinthians 1:20-22
For no matter how many promises God has made, they are "Yes" in Christ. And so through him, the "Amen" is spoken by us to the glory of God. Now it is God who makes both us and you stand firm in Christ. He anointed us, set his seal of ownership on us, and put his Spirit in our hearts as a deposit, guaranteeing what is to come.

Ephesians 3:20
Now to him who is able to do immeasurably more than all we ask or imagine, according to his power that is at work within us.

Hebrews 10:39
But we are not of those who shrink back and are destroyed, but of those who believe and are saved.

2 Peter 3:9-10
The Lord is not slow in keeping his promise, as some understand slowness. He is patient with you, not wanting anyone to perish, but everyone to come to repentance.

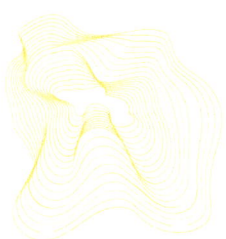

9 The Periodic Table

I saw in a dream a table where all the elements fell into place as required. Awakening, I immediately wrote it down on a piece of paper.
 - Dmitri Mendeleev, Russian chemist

As early as the late 1700s, scientists have been discovering and organizing the elements found in our universe. (An element is two or more atoms with the same number of protons.) The current periodic table is organized by atomic numbers, which are equal to the number of protons in each element. The numbers are in ascending order, moving from left to right across the table. Through this organizational system, the groups and rows tell us specific things about the elements.

For example, Calcium is atomic number 20. Calcium is in group 2 and row 4. This placement tells me that calcium has 20 protons, it is an alkaline-earth metal, and it has four energy levels, as well as two electrons in its fourth energy level. Elements in the same group as Calcium share similar physical and chemical properties.

The scientists who worked to perfect the organization system of the periodic table tapped into one of the most important traits of God's nature. In 1 Corinthians 14:33, Paul tells us, "For God is not a God of disorder but of peace." Our God is not the god of confusion or disarray.

We serve a God of order, wisdom, and peace. When we find ourselves confused or in a place of strife, we can remind ourselves of whom we serve and allow Him to sort it all out.

1 Corinthians 14:40
But everything should be done in a fitting and orderly way.

Colossians 2:5
For though I am absent from you in body, I am present with you in spirit, and delight to see how orderly you are and how firm your faith in Christ is.

2 Thessalonians 3:16
Now may the Lord of peace himself give you peace at all times and in every way. The Lord be with all of you.

James 3:17
But the wisdom that comes from heaven is first of all pure; then peaceloving, considerate, submissive, full of mercy and good fruit, impartial and sincere.

Revelation 21:4
He will wipe every tear from their eyes. There will be no more death or mourning or crying or pain, for the old order of things has passed away.

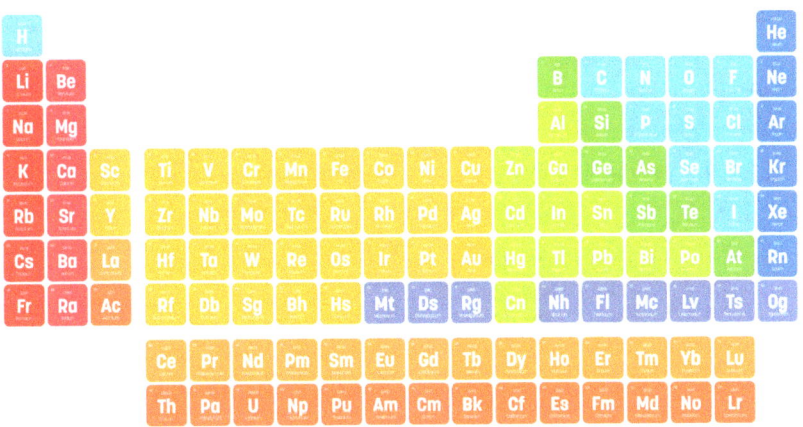

10 Noble Gases

"What we know is a drop, what we don't know is an ocean."
 - Isaac Newton, English polymath

Noble gases are a unique group of elements listed on the periodic table. Group 18 is known for its high stability and low chemical reactivity. The reason for these characteristics is due to the fact that these gases have an outer energy level that has enough electrons to make it full.

All other elements have an outer energy level that is missing one or more of the desired eight electrons. As a child of God, there are times we feel we are missing something. It usually happens when we forget who we are in Christ. When we fail to remember all God has provided for us, we begin to feel unstable and act in ways that don't reflect our true selves.

In Luke 15, we find the story of The Prodigal Son. The younger of the two sons wanted his inheritance and set out on his way. He traveled to a distant land and wasted his money. He began to be in need, lacking all the things he had enjoyed at home. He searched for help and ended up feeding pigs.

He was homeless, starving, and dirty. When he came to himself, he decided to return home. The father was waiting and had the servants bring out the best robe to put on him, a ring to put on his finger, and shoes for his feet. The prodigal was restored to all the rights and privileges of a son.

Like the prodigal, we are sons and daughters of nobility, dressed in robes of righteousness with a ring on our finger, a crown on our head, and cared for by our loving Father, the King of kings.

Isaiah 61:10
I delight greatly in the Lord; my soul rejoices in my God. For he has clothed me with garments of salvation and arrayed me in a robe of righteousness, as a bridegroom adorns his head like a priest, and as a bride adorns herself with her jewels.

Acts 17:11
Now the Bereans were of more noble character than the Thessalonians, for they received the message with great eagerness and examined the Scriptures every day to see if what Paul said was true.

Luke 15:22-24
"But the father said to his servants, 'Quick! Bring the best robe and put it on him. Put a ring on his finger and sandals on his feet. Bring the fattened calf and kill it. Let's have a feast and celebrate. For this son of mine was dead and is alive again; he was lost and is found.' So they began to celebrate."

1 Peter 2:9
But you are a chosen people, a royal priesthood, a holy nation, a people belonging to God, that you may declare the praises of him who called you out of darkness into his wonderful light.

Revelation 7:9
After this, I looked and there before me was a great multitude that no one could count; from every nation, tribe, people, and language, standing before the throne and in front of the Lamb. They were wearing white robes and were holding palm branches in their hands.

11 Chemical Bonds

He who thinks half-heartedly will not believe in God, but he who really thinks has to believe in God. - Isaac Newton, English polymath

Everything in our physical world is a combination of the elements listed on the periodic table. The elements can be divided into three categories, called families. These families are metals, metalloids, and nonmetals. All elements bond with like elements or a combination of one or more different elements. The bonding occurs in three different ways: metal to metal, metal to nonmetal, and nonmetal to nonmetal.

When elements bond, the original properties and characteristics of the elements change. For example, the element sodium is highly reactive with water. The element chlorine is an extremely poisonous substance.   When sodium and chlorine form a chemical bond, they create table salt. Table salt, chemically known as sodium chloride, is generally considered unreactive and safe to consume and is commonly used as a food preservative.

A remarkable transformation occurs when we bond with Christ. When we choose to receive the gift of salvation, we leave our old, sinful life behind and step into the newness of life that Jesus died to provide for us.

In 2 Corinthians 5:17, we find a beautiful reminder: Therefore, if anyone is in Christ, he is a new creation; the old has gone, the new has come.

1 Samuel 10:6-7

The Spirit of the Lord will come upon you in power, and you will prophesy with them; and you will be changed into a different person. Once these signs are fulfilled, do whatever your hand finds to do, for God is with you.

Romans 6:4

We were therefore buried with him through baptism into death in order that, just as Christ was raised from the dead through the glory of the Father, we too may live a new life.

Galatians 6:15

Neither circumcision nor uncircumcision means anything; what counts is a new creation.

Ephesians 4:22-24

You were taught, with regard to your former way of life, to put off your old self, which is being corrupted by its deceitful desires; to be made new in the attitude of your minds; and to put on the new self, created to be like God in true righteousness and holiness.

Colossians 3:9-10

Do not lie to each other, since you have taken off your old self with its practices and have put on the new self, which is being renewed in knowledge in the image of its Creator.

12 Ionic Bonds

The name's Bond, Ionic Bond; taken, not shared.
- unknown

An ionic bond is a type of chemical bond in which a metal element bonds with a nonmetal element. These different types of elements are drawn together because each has something the other element needs. They both have an outer energy level that is not considered full. But when these two types of opposite elements bond, they transfer the needed electrons from one to the other and create a compound that satisfies both.

Potassium is a metal. It has 19 total electrons divided among 4 energy levels. The first energy level holds two electrons, the second and third energy levels hold eight electrons each, and the fourth energy level needs to hold eight to be full, but it only has one. Potassium wants a full outer energy level, so the easiest thing to do is to give the one electron to an element that needs it.

An element that frequently bonds with potassium is the nonmetal fluorine. Fluorine has 9 total electrons and two energy levels. This means the outermost level has seven electrons and only needs one more to be full. Potassium fluoride is the compound created by potassium giving its one electron to fluorine, and fluorine taking the extra electron from potassium. Ionic bonding is a give-and-take, or a transfer from one atom to another.

In Acts 20:35, Paul quotes Jesus, "It is more blessed to give than to receive." God is a generous, giving Father, and we are to imitate Him. Taking comes naturally to the human part of all of us, but as His followers, we have a new nature and we must daily choose to be a giver.

Can you think of something you can give away today?

Deuteronomy 15:10-11
Give generously to him and do so without a grudging heart; then, because of this, the Lord your God will bless you in all your work and in everything you put your hand to. There will always be poor people in the land. Therefore, I command you to be open-handed toward your brothers and toward the poor and needy in your land.

Proverbs 11:25
A generous man will prosper; he who refreshes others will be refreshed.

Malachi 3:10
"Bring the whole tithe into the storehouse, that there may be food in my house. Test me in this," says the Lord Almighty, "see if I will not throw open the floodgates of heaven and pour out so much blessing that you will not have room enough for it."

Luke 6:38
"Give, and it will be given to you. A good measure, pressed down, shaken together, and running over, will be poured into your lap. For with the measure you use, it will be measured to you."

2 Corinthians 9:7
Each man should give what he has decided in his heart to give, not reluctantly or under compulsion, for God loves a cheerful giver.

13 Cations and Anions

I have come to believe that giving and receiving are the same. Giving and receiving - not giving and taking. - Joyce Grenfell, English singer and actress

Cations and anions are names given to the types of ions created in Ionic Bonds. Ions are charged particles. When the proton and electron amounts are equal, the whole atom is neutral, meaning the atom has no charge. When atoms transfer electrons to one another, the number of electrons is different from the number of protons in each element. When these amounts are unequal, the atoms become charged.

Cations are positively charged, and anions are negatively charged. When an atom gives away its excess electrons, the atom will then have more protons, which creates an overall positive charge. When an atom gains electrons, causing it to have more electrons than protons, the atom will have a negative charge overall.

In our day-to-day interactions, we must learn how to have a positive impact on others and take on some of the negative challenges they may be facing. In Galatians 6:2, we are told to carry each other's burdens. Giving our burdens away or taking burdens from others is not easy for most of us.

For some, it's hard to allow others to help us when we are overwhelmed or in need of help. It can also be difficult for us to step up and be there for others when they need a burden lifted.

Choosing to let others help or being the one willing to let someone help is one of the practical ways we show Christ to others. If there is a burden you need to transfer to a friend or you are the friend who needs to take a burden away, don't delay any longer.

Proverbs 19:17
He who is kind to the poor lends to the Lord, and he will reward him for what he has done.

Matthew 7:12
So in everything, do to others what you would have them do to you, for this sums up the Law and the Prophets.

Romans 12:13
Share with God's people who are in need. Practice hospitality.

Ephesians 4:32
Be kind and compassionate to one another, forgiving each other, just as in Christ God forgave you.

Hebrews 13:16
And do not forget to do good and to share with others, for with such sacrifices God is pleased.

14 Covalent Bonds

The nature of the chemical bond is the problem at the heart of all chemistry.
- Bryce Crawford, American street evangelist

A covalent bond forms when a nonmetal element bonds with a nonmetal element. These types of bonds form when the atoms share electrons. These shared bonds are stronger than other kinds of bonds.

Another characteristic of these types of bonds is that they are directional. This means the atoms bond in specific orientations in space. Take water, for example, $H2O$, which is a covalent bond where the oxygen atom is always in the center with one hydrogen atom on each side, creating its unique shape.

From this type of chemical bond, I am reminded of the word covenant. A covenant is a relationship between two partners who make a binding promise to one another and work together to reach a common goal.

Covenants are relational and personal. The partners share in creating a strong bond. Jesus became human and shared in our humanity so that we could share in all He is preparing for us. When we choose to come into a relationship with Jesus, all that we are and all that we have become His, and all that He is and has becomes ours.

What an amazing thing to think about! By one sacrifice, He has made perfect forever those who are being made holy (Hebrews 10:14); you and me!

Jeremiah 31:33b-34
"I will put my law in their minds and write it on their hearts. I will be their God, and they will be my people. No longer will a man teach his neighbor, or a man his brother. Saying, 'Know the Lord,' because they will all know me, from the least of them to the greatest," declares the Lord.

2 Corinthians 3:6
He has made us competent as ministers of a new covenant, not of the letter but of the Spirit; for the letter kills, but the Spirit gives life.

Colossians 1:12
Giving thanks to the Father, who has qualified you to share in the inheritance of the saints in the kingdom of light.

2 Thessalonians 2:13-14
But we ought always to thank God for you, brothers loved by the Lord, because from the beginning God chose you to be saved through the sanctifying work of the Spirit and through belief in truth. He called you to this through our gospel, that you might share in the glory of the Lord Jesus Christ.

Hebrews 9:15
For this reason, Christ is the mediator of a new covenant, that those who are called may receive the promised eternal inheritance, now that he has died as a ransom to set them free from the sins committed under the first covenant.

15 Triple Bonds

When the chemistry is right, all the experiments work.
 - Gregory Benford, American astrophysicist

There are different types of covalent bonds. One type is called a triple bond. A triple bond occurs when atoms share three pairs of electrons, totaling six bonding electrons. This triple sharing creates the strongest type of covalent bond because the shared pairs pull the electrons closer to the nucleus. These bonds are very hard to break apart.

A common example of a triple bond is Nitrogen (N_2). Two nitrogen atoms will bond together because each atom has only five electrons in its outermost level and needs three more to fill the outer level. Each atom shares three of its electrons with the other and creates a triple bond.

Have you ever noticed how many times the number three is used in Scripture? Here are a few examples: In the Old Testament, Abraham, Isaac, and Jacob are foundational figures, and Jonah was in the belly of the fish for three days. In the New Testament, Jesus' ministry lasted three years, and He was resurrected on the third day, but the most prominent group of three is the Trinity: Father, Son, and Holy Spirit.

In the bible, the number three represents completeness and perfection. Each day in our walk with the Lord, we won't easily be 'broken' if we remember these three virtues found in 1 Corinthians 13:13: 'And now these three remain: faith, hope, and love. But the greatest of these is love.'

Romans 10:17
Consequently, faith comes from hearing the message, and the message is heard through the word of Christ.

2 Corinthians 5:14-15
For Christ's love compels us, because we are convinced that one died for all, and therefore all died. And he died for all, that those who live should no longer live for themselves but for him who died for them and was raised again.

Colossians 1:22-23
But now he has reconciled you by Christ's physical body through death to present you holy in his sight, without blemish and free from accusation— If you continue in your faith, established and firm, not moved from the hope held out in the gospel.

1 Thessalonians 1:3
We continually remember before our God and Father your work produced by faith, your labor prompted by love, and your endurance inspired by hope in our Lord Jesus Christ.

Hebrews 11:1
Now faith is being sure of what we hope for and certain of what we do not see.

16 Naming Compounds

The meeting of two personalities is like the contact of two chemical substances; if there is any reaction, both are transformed. - Carl Jung, Swiss psychiatrist

In chemistry, when compounds are created by chemical bonding, they are given names based on a few basic rules. If the compound is created by an ionic bond, the chemicals involved keep their names with the metal element's name listed first, followed by the nonmetal's name. The nonmetal element's name gets the suffix *-ide* added to it. For example, if sodium, a metal, and iodine, a nonmetal, bond, the new name would be sodium iodide.

If the compound is created by a covalent bond, the elements involved in the bonding follow a few more rules when given a name. First, the elements will be listed in a specific order based on their position on the periodic table. Then, all the elements will receive a prefix indicating how many atoms of the element are in the molecule. Lastly, the last element listed in the name will also receive a suffix. For example, the chemical name of water is dihydrogen monoxide. Hydrogen is listed first because it is one row higher on the periodic table than oxygen. The prefix *di-* is added to hydrogen because there are two hydrogen atoms, and *mono-* is added to oxygen because there is only one oxygen atom. Finally, the suffix *-ide* is also added to oxygen.

In the Bible, names are important. We can learn a lot about a person, their personality, character, and how God used them to fulfill His plan. In Genesis, Abram becomes Abraham. Why would God change his name? Abram means "exalted father", but Abraham means "father of many nations". God made a covenant with Abram, and his new name revealed the plan this covenant would fulfill.

When we enter into a relationship with God, we have the promise of being given a new name, too. In Revelation 2:17, we read, "To him who overcomes…I will also give him a white stone with a new name written on it, known only to him who receives it.

Genesis 17:5

No longer will you be called Abram; your name will be Abraham, for I have made you a father of many nations.

Genesis 35:10-12

God said to him," Your name is Jacob, but you will no longer be called Jacob; your name will be Israel." So he named him Israel. And God said to him, " I am God Almighty, be fruitful and increase in number. A nation and a community of nations will come from your body. The land I gave to Abraham and Isaac, I also give to you, and I will give this land to your descendants after you."

Isaiah 62:2

The nations will see your righteousness, and all kings your glory; you will be called by a new name that the mouth of the Lord will bestow.

Matthew 16:17-18

Jesus replied, "Blessed are you, Simon son of Jonah, for this was not revealed to you by man, but my Father in heaven. And I tell you that you are Peter, and on this rock I will build my church, and the gates of Hades will not overcome it.

Acts 13:9

Then Saul, who was also called Paul, filled with the Holy Spirit...

17 Pure Substances

Silver is purified in fire, and so are we. It is in the most trying times that our real character is shaped and revealed. - Helen Keller, American author

In the language of chemistry, a pure substance is defined as a material composed of only one type of atom or molecule with a fixed ratio. A pure substance cannot be separated by physical means. Each substance has its own unique chemical and physical properties.

Some examples of pure substances are water, gold, silver, salt, and even air. Elements like gold and silver have been mined for thousands of years. Elements like these are found in ore, a rock that contains valuable elements or compounds. To get the pure substance out of the ore, it must be refined. A metalsmith heats the ore until it melts and dross begins to accumulate on top of the melted material. Dross refers to the impurities or unwanted materials that are removed during the refining process. If the scum or impurities remained, the pure substances would be considered defective or contaminated.

As Christians in a fallen world, we are continually being refined because our lives are still full of impurities that God calls sin. In Malachi 3:2, we find a promise that Jesus will be like a refiner's fire or a launderer's soap. When Jesus shed His blood, that promise was fulfilled. The blood of Jesus purifies us. Our spirit man is washed and cleansed the moment we ask Jesus into our lives, but our mind and flesh still contain the habits of our sinful nature.

The refining process is uncomfortable and sometimes requires pain, hardship, suffering, and sorrow, but we can endure when we know who the Refiner is and the good plan that He has in store for us.

Malachi 3:2-3
But who can endure the day of his coming? Who can stand when he appears? For he will be like a refiner's fire or a launderer's soap. He will sit as a refiner and purifier of silver; he will purify the Levites and refine them like gold and silver.

2 Timothy 2:3-4
Endure hardship with us like a good soldier of Christ Jesus. No one serving as a soldier gets involved in civilian affairs – he wants to please his commanding officer.

2 Timothy 2:22
Flee the evil desires of youth, pursue righteousness, faith, love, and peace along with those who call on the Lord out of a pure heart.

Hebrews 12:10-11
Our fathers disciplined us for a little while as they thought best, but God disciplines us for our good, that we may share in his holiness. No discipline seems pleasant at the time, but painful. Later on, however, it produces a harvest of righteousness and peace for those who have been trained by it.

1 Peter 1:6-7
In this, you greatly rejoice, though now for a little while you may have had to suffer grief in all kinds of trials. These have come so that your faith—of greater worth than gold, which perishes even though refined by fire—may be proved genuine and may result in praise, glory, and honor when Jesus Christ is revealed.

18 Silver

Refining is inevitable in science when you have made measurements of a phenomenon for a long period of time. - Charles Richter, American physicist

Silver has been a source of power and wealth since ancient times. Silver has been used as currency in the form of coins. It has and still is used in jewelry and household items. It was also used for medicinal purposes due to the belief that it has antibacterial properties. The Egyptians believed it had properties for purifying water and considered silver more valuable than gold because it was rarer.

Silver is number 47 on the periodic table. Its chemical symbol is Ag, coming from the Latin word for argentum, which means 'silver' or 'shiny'. Due to its lustrous property, silver is the best reflector of all the elements on the periodic table.

Like silver, we are to shine and reflect Jesus. Matthew 5:16 says, "Let your light shine before men that they may see your good deeds and praise your Father in heaven."

We are called to live openly and visibly so others can see our faith in action. When we perform acts of kindness, walk in love towards others, and live a holy, set-apart life, we are pointing others toward God. We are reflecting His light, and others see Him in us.

Matthew 25:1-4
At that time the kingdom of heaven will be like ten virgins who took their lamps and went out to meet the bridegroom. Five of them were foolish and five were wise. The foolish ones took their lamps but did not take any oil with them. The wise, however, took oil in jars along with their lamps.

Luke 8:16
No one lights a lamp and hides it in a jar or puts it under a bed. Instead, he puts it on a stand, so that those who come in can see the light.

Acts 13:47
For this is what the Lord has commanded us: "I have made you a light for the Gentiles, that you may bring salvation to the ends of the earth."

2 Corinthians 3:18
And we, who with unveiled faces all reflect the Lord's glory, are being transformed into his likeness with ever-increasing glory, which comes from the Lord, who is the Spirit.

1 John 2:9-11
Anyone who claims to be in the light but hates his brother is still in the darkness. Whoever loves his brother lives in the light, and there is nothing in him to make him stumble. But whoever hates his brother is in the darkness; he does not know where he is going, because the darkness has blinded him.

19 Gold

You are an alchemist; make gold of that.
- Shakespeare, English playwright

Gold, element 79. Gold is the most malleable and ductile of all metals. It is also one of the softest and heaviest metals. Gold is an excellent conductor of heat and electricity, and it is resistant to attack by air, heat, moisture, and most solvents. Because of gold's rarity, durability, color, and chemical properties, gold has been valuable throughout history.

Gold is first mentioned in Genesis chapter two. No other metal has been more frequently mentioned in the Old Testament than gold. Even into the New Testament, having gold was the most convenient way of storing up wealth. It has always been easy for men, even godly men, to store up wealth; some even go to great lengths to gain it and keep it safe.

In Matthew 6:19-21, Jesus said, "Do not store up for yourselves treasures on earth, where moth and rust destroy and where thieves break in and steal. But store up for yourselves treasures in heaven, where moth and rust do not destroy, and where thieves do not break in and steal. For where your treasure is, there your heart will be also." Jesus even tells us in other parts of the Gospels that we "store up treasure" in heaven by sharing His love with others and helping those in need.

Just think, one day when we arrive in heaven, gold will be so abundant and common that we will walk on streets made of it, and the treasures we've sent ahead of us will be safely waiting for us to enjoy throughout all of eternity.

Matthew 10:42
"And if anyone gives even a cup of cold water to one of these little ones because he is my disciple, I tell you the truth, he will certainly not lose his reward."

Luke 12:33
Sell your possessions and give to the poor. Provide purses for yourselves that will not wear out, a treasure in heaven that will not be exhausted, where no thief comes near and no moth destroys.

Luke 18:22
When Jesus heard this, he said to him, "You still lack one thing. Sell everything you have and give to the poor, and you will have treasure in heaven. Then come, follow me."

2 Timothy 4:8
Now there is in store for me the crown of righteousness, which the Lord, the righteous Judge, will award to me on that day— and not only to me, but also to all who have longed for his appearing.

Revelation 22:12
"Behold, I am coming soon! My reward is with me, and I will give to everyone according to what he has done."

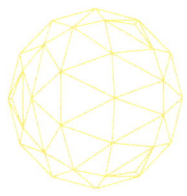

20 Mixtures

What is the most important thing to learn in chemistry? Never lick the spoon.
- unknown

Mixtures are substances that can be combined and/or separated by physical means. Some mixtures combine and blend so well that they look uniform throughout, like mayonnaise or air. Other substances are mixed, but the different parts that make up the mixture are still easily seen, like trail mix snacks or chocolate chip cookies.

When physically combining substances, stirring is one of the crucial methods used. Stirring ensures everything is evenly mixed throughout the substance. Stirring helps distribute heat evenly and speeds up the dissolving of solids in liquids.

Imagine a glass of sweet tea. When sugar is added, it will fall straight to the bottom and rest there until stirred. Stirring ensures the sugar is fully incorporated so the entire beverage is sweetened. In 2 Timothy 1:6, as Timothy begins his ministry, Paul reminds him to stir up the gift of God within him.

As God's children, we have been given spiritual gifts to equip us for godly service, but in the beginning of our Christian walk, these gifts are undeveloped.

Through practice and perseverance, we learn by actively allowing the Holy Spirit to show us how to use them properly. Stirring up spiritual gifts keeps us active and effective in our service to the Lord.

Job 17:9
Nevertheless, the righteous will hold to their ways, and those with clean hands will grow stronger.

Romans 12:6-8
We have different gifts, according to the grace given us. If a man's gift is prophesying, let him use it in proportion to his faith. If it is serving, let him serve; if it is teaching, let him teach; if it is encouraging, let him encourage; if it is contributing to the needs of others, let him give generously; if it is leadership, let him govern diligently; if it is showing mercy, let him do it cheerfully.

1 Corinthians 12:8-11
To one there is given through the Spirit the message of wisdom, to another the message of knowledge by means of the same Spirit, to another faith by the same Spirit, to another gifts of healing by that one Spirit, to another miraculous powers, to another speaking in different kinds of tongues, and to still another the interpretation of tongues. All these are the work of one and the same Spirit, and he gives them to each one, just as he determines.

Galatians 5:22-23
But the fruit of the Spirit is love, joy, peace, patience, kindness, goodness, faithfulness, gentleness, and self-control. Against such things there is no law.

2 Timothy 1:6-7
For this reason I remind you to fan into flame the gift of God, which is in you through the laying on of my hands. For God did not give us a spirit of timidity, but a spirit of power, of love, and of self-discipline.

21 Alloys

If Iron Man and Silver Surfer teamed up, they'd be alloys.
- unknown

Alloys are purposefully engineered chemical compounds designed to enhance the properties of materials used in everyday items. Alloyed materials become much stronger and more durable than the original.

Alloying certain metals can increase their hardness, making them more resistant to scratches and wear. Some alloys are designed to be resistant to corrosion, while others are created for specific applications like a stronger but lighter baseball bat or golf club.

In our Christian Walk, God often places other people in our lives who are intended to make us stronger, to help us become ready for the specific task He wants us to complete. God will bring older and wiser people into our lives so that we can learn from their experience; conversely, we could be the ones sent to give our expertise to someone in need of growth or strength.

In Titus 2, Paul tells Timothy to teach older men and to encourage older women to train younger women. In a community of diverse ages and skills, we can learn a lot from one another. Isn't it just like a loving God to engineer various relationships to provide just the mix of people and personalities we need to make each other better as we walk this life together?

Psalms 133:1
How good and pleasant it is when brothers live together in unity!

Proverbs 27:9
Perfume and incense bring joy to the heart, and the pleasantness of one's friend springs from his earnest counsel.

Romans 12:10
Be devoted to one another in brotherly love. Honor one another above yourselves.

Titus 2:2-6
Teach the older men to be temperate, worthy of respect, self-controlled, and sound in faith, in love, and in endurance. Likewise, teach the older women to be reverent in the way they live, not to be slanderers or addicted to too much wine, but to teach what is good. Then they can train the younger women to love their husbands and children, to be self-controlled and pure, to be busy at home, to be kind, and to be subject to their husbands, so that no one will malign the word of God. Similarly, encourage the young men to be self-controlled.

1 Peter 4:8-10
Above all, love each other deeply, because love covers over a multitude of sins. Offer hospitality to one another without grumbling. Each one should use whatever gift he has received to serve others, faithfully administering God's grace in its various forms.

22 Biochemistry

To know that we know what we know and to know that we do know what we do not know, that is true knowledge. - Nicolaus Copernicus, Astronomer

The prefix *-bio* means life. When added to the term chemistry, we have the branch of science that studies the chemical processes within and surrounding living things, known as Biochemistry.

Biochemistry is central to understanding DNA, the molecule that carries the blueprint for all living organisms. Every cell contains the information each organism needs to grow and develop, respond to its environment, breathe, reproduce, and use energy. Our unique blueprint makes us who we are, and nothing about us was an accident or just evolutionary chance.

In Psalms 139, we find a beautiful reminder of the care and plan God had for us from the very beginning. King David, inspired by the Holy Spirit, reminds us that God created our inmost being. He knit us together in our mother's womb. We are fearfully and wonderfully made. Our frame was not hidden from God when we were made in the secret place. All the days ordained for us were written in His book before one of them came to be.

Think about that! We are uniquely crafted, and no part of us: how we look, behave, the things we enjoy, and even the timing of our birth, is a mistake. We are His handiwork. We have a purpose, and we are so very loved.

Genesis 1:27
So God created man in his own image, in the image of God he created him; male and female he created them.

Joshua 1:9
Have I not commanded you? Be strong and courageous. Do not be terrified; do not be discouraged, for the Lord your God will be with you wherever you go.

Psalms 139:13-16
For you created my inmost being; you knit me together in my mother's womb. I praise you because I am fearfully and wonderfully made; your works are wonderful, I know that full well. My frame was not hidden from you when I was made in the secret place. When I was woven together in the depths of the earth, your eyes saw my unformed body. All the days ordained for me were written in your book before one of them came to be.

Jeremiah 1:5
"Before I formed you in the womb I knew you; before you were born I set you apart; I appointed you as a prophet to the nations."

Ephesians 2:10
For we are God's workmanship, created in Christ Jesus to do good works, which God prepared in advance for us to do.

23 Carbon

I have no fear of losing my life - If I have to save a koala or a crocodile or a kangaroo or a snake, mate, I will save it. - Steve Irwin, Australian conservationist

Carbon is element number 6 on the periodic table. It has four electrons in its outermost energy level, allowing it to bond with other elements in a variety of ways. Carbon forms more compounds than all the other elements combined. Carbon is essential for life. It is constantly exchanged between living and nonliving processes.

The Carbon Cycle is an explanation of the continuous exchange of carbon atoms between the atmosphere, the Earth's surface, and living organisms. In today's society, we often hear conversations about a term called the 'carbon footprint'. Scientists believe that humans harm the carbon cycle through various human-engineered mechanisms that create pollution and waste.

From the beginning in Genesis 2:15, we find that Adam's first job was to work the ground and take care of what God had entrusted to him. When Adam sinned and the fall of mankind spread, the whole earth groaned - and in Romans 8, we are told that the creation is eagerly waiting for the world to be redeemed so that it can be freed from its bondage of decay.

As we walk this earth, may our footprints leave only traces of care, concern, and compassion. May we be good stewards over all of His creation until He returns.

Psalms 24:1-2
The earth is the Lord's and everything in it, the world, and all who live in it; for he founded it upon the seas and established it upon the waters.

Psalms 89:11
The heavens are yours, and yours also the earth; you founded the world and all that is in it.

Psalms 104:24-27
How many are your works, O Lord! In wisdom you made them all; the earth is full of your creatures. There is the sea, vast and spacious, teeming with creatures beyond number—living things both large and small. The ships go to and fro, and the leviathan, which you formed to frolic there. These all look to you to give them food at the proper time.

Romans 8:19-21
The creation waits in eager expectation for the sons of God to be revealed. For the creation was subjected to frustration, not by its own choice, but by the will of the one who subjected it, in hope that the creation itself will be liberated from its bondage to decay and brought into glorious freedom of the children of God.

1 Corinthians 4:2
Now it is required that those who have been given a trust must prove faithful.

24 pH Levels

If the human body is balanced in pH and nutrients, it is not susceptible to disease.
- Royal Rife, American inventor

pH is a measure of how acidic or basic a substance is. pH is measured on a scale of 0 to 14. When the pH value is 7, the substance is considered neutral, which means it is neither acidic nor basic. A pH value of less than 7 means a substance has a high concentration of hydrogen ions and is considered an acid. Substances that register above 7 are considered alkaline or basic because they have high concentrations of hydroxide ions.

Acidic substances are sour, corrosive, reactive with metals, and dissolve in water. Bases, on the other hand, have a slippery feel, conduct electricity in water, and promote certain chemical reactions.

It may surprise you to know that there are examples of pH in the Bible: sour grapes, wine, and vinegar; leaven and fermentation; butter, water, herbs, and soil quality are just a few. Jesus teaches about soil quality in Matthew 13: The Parable of the Sower.

This parable focuses on the seed, a symbol of the Word of God, and types of soil, which symbolize our hearts. Similar to a real garden being sown, if the soil is too acidic or too basic, the seed won't get the nutrients it needs.

Jesus explains this in Matthew 13:23. "But the one who received the seed that fell on good soil is the man who hears the Word and understands it."

In the garden of our heart, we need to mind our pH as well.

Mark 4:20
Others, like the seed sown on good soil, hear the word, accept it, and produce a crop- thirty, sixty, or even a hundred times what was sown.

Luke 8:15
But the seed on good soil stands for those with a noble and good heart, who hear the word, retain it, and by persevering produce a crop.

Luke 11:28
He replied, "Blessed rather are those who hear the word of God and obey it."

2 Corinthians 9:10
Surely he says this for us, doesn't he? Yes, this was written for us, because when the plowman plows and the thresher threshes, they ought to do so in hope of sharing in the harvest.

James 1:22
Do not merely listen to the word, and so deceive yourselves. Do what it says.

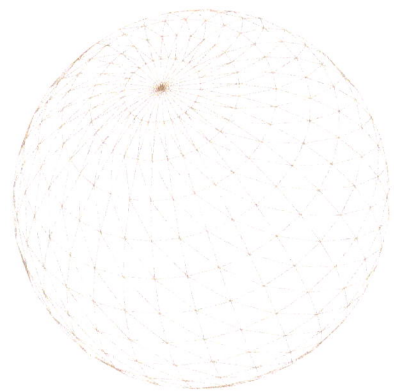

25 Blood

Blood is the life-maintaining fluid that flows through the body's blood vessels. Blood carries nourishment, electrolytes, hormones, vitamins, antibodies, heat, and oxygen to all body tissues. Blood also takes waste and carbon dioxide away.

Blood cannot be expressed with a single, simple formula like that of a pure compound. Blood's purpose, however, can be simplified by dividing its jobs into three categories: transportation, protection, and regulation.

Human blood has limitations on the life it provides for us, but there is another Blood that, when applied to our life, has unlimited power, washes us white as snow, and provides eternal life: The Blood of Jesus. Jesus' blood redeems us, covers us, cleanses us, justifies us, sets us apart, protects, and delivers us.

Because Jesus chose to endure the cross and shed his blood for us, He made the way for us to live forever with Him. His blood gives us victory over evil, His blood sanctifies and purifies, making us holy in the Father's sight. Colossians 1:22 declares, 'But **now** he has reconciled you by Christ's physical body through death to present you holy in his sight, without blemish and free from accusation.'

Thank you, Jesus, for the blood!

Ephesians 1:7-8
In him we have redemption through his blood, the forgiveness of sins, in accordance with the riches of God's grace that he lavished on us with all wisdom and understanding.

Colossians 1:19-20
For God was pleased to have all his fullness dwell in him, and through him to reconcile to himself all things, whether things on earth or things in heaven, by making peace through his blood shed on the cross.

Hebrews 9:28
So Christ was sacrificed once to take away the sins of many people; and he will appear a second time, not to bear sin, but to bring salvation to those who are waiting for him.

1 John 1:9
If we confess our sins, he is faithful and just and will forgive us our sins and purify us from all unrighteousness.

Revelation 12:11
They overcame him by the blood of the Lamb and by the word of their testimony; they did not love their lives so much as to shrink from death.

26 Water

The universal solvent: water. Water is essential to the life of living things. It can dissolve more substances than any other liquid. It is also the only substance that naturally exists in three states: solid (ice), liquid, and gas (steam). Water is a transparent, odorless, and nearly colorless liquid at room temperature.

Water has a unique density. Most substances become denser when they are in a solid state, but solid water becomes less dense than liquid water. This unusual property allows ice to float on liquid water, creating an insulating layer on lakes and oceans. This insulation protects aquatic life from freezing in winter. In the bible, we can find Jesus sharing an unusual property of the water from heaven.

In John 4, Jesus meets a Samaritan woman at a well and asks her to give Him a drink of water. She is surprised because usually Jews did not associate with Samaritans. He proceeds to tell her that the water from the well will only quench thirst for a while, but whoever drinks from the water He gives will never thirst. Jesus was giving a preview of the indwelling of the Holy Spirit.

The Holy Spirit is frequently symbolized as water in scripture. Jesus says in John 7:38, "whoever believes in me, as the scripture has said, streams of living water will flow from within him."

Ezekial 36:25-27
I will sprinkle clean water on you, and you will be clean; I will cleanse you from all your impurities and all your idols. I will give you a new heart and put a new spirit in you; I will remove from you your heart of stone and give you a heart of flesh. And I will put my Spirit in you and move you to follow my decrees and be careful to keep my laws.

Joel 2:28-29
"And afterward, I will pour out my Spirit on all people. Your sons and daughters will prophesy, your old men will dream dreams, your young men will see visions. Even on my servants, both men and women, I will pour out my Spirit in those days."

Zechariah 12:10
"And I will pour out on the house of David and the inhabitants of Jerusalem a spirit of grace and supplication. They will look on me, the one they have pierced, and they will mourn for him as one mourns for an only child, and grieve bitterly for him as one grieves for a firstborn son.

John 4:13-15
Jesus answered, "Everyone who drinks this water will be thirsty again, but whoever drinks the water I give him will never thirst. Indeed, the water I give him will become in him a spring of water welling up to eternal life."

John 7:37-38
On the last and greatest day of the Feast, Jesus stood and said in a loud voice, "If anyone is thirsty, let him come to me and drink. Whoever believes in me, as the scripture has said, streams of living water will flow from within him."

Part 2
Physics

Physics is the science that studies matter and energy, and how they interact and behave in space and time.

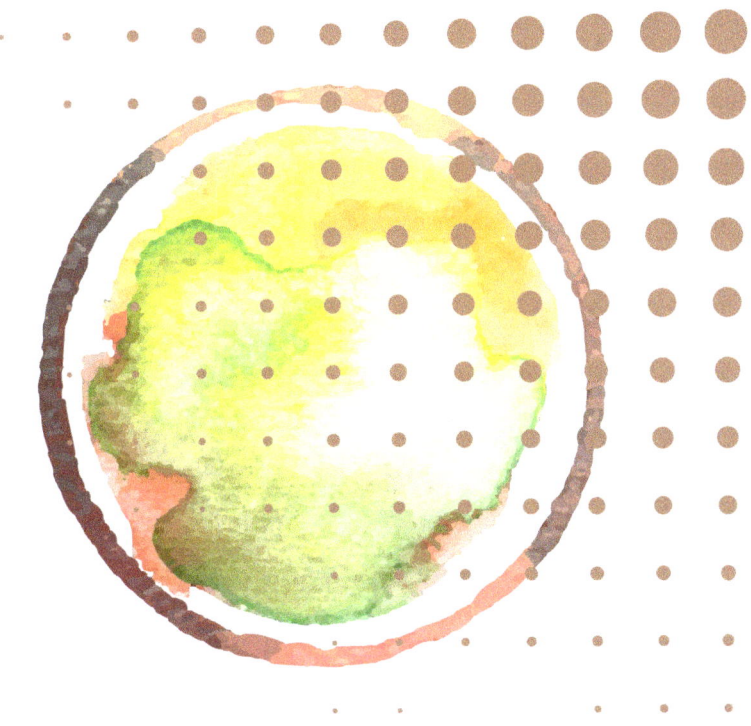

27　Kinematics

All of physics is either impossible or trivial. It is impossible until you understand it, and then it becomes trivial. - Ernest Rutherford, New Zealand physicist

Kinematics is the branch of mechanics that focuses on motion. Kinematics studies the movement of objects without worrying about what makes them move. Kinematics describes relationships between position, velocity, and acceleration using mathematical equations and principles. To begin to understand kinematics, you have to start by understanding motion.

Motion is simply the change in the position of an object over time. There are several types of motion: Circular motion, like a Ferris wheel, or rotational motion, like a spinning top. Back-and-forth motion, like a vibrating string on a guitar, demonstrates oscillatory motion, and random motion is anything without a predictable pattern, like insects flying about.

From the moment we wake until the second our head hits the pillow, we have spent hours engaged in many types of motion. We change our position or place constantly, and it's exhausting. God has given us a way to move through each day without growing tired. Isaiah 40:31 tells us, "Those who wait on the Lord will renew their strength."

To wait doesn't necessarily mean we have to cease our moving; it means that in all our motion, we move with our heart and mind staying expectant, hopeful, and patient. It means we move in step with the Lord.

We move, expecting that He orders our steps. We move, hopeful that our needs will be met. We move, waiting patiently for our prayers to be answered by Him and not manipulated by our means. Waiting is trusting that He will get us where we need to be and help us move as He directs.

Exodus 14:14
The Lord will fight for you; you need only to be still.

Psalms 27:14
Wait for the Lord; be strong and take heart and wait for the Lord.

Psalms 130:5-6
I wait for the Lord, my soul waits, and in his word I put my hope. My soul waits for the Lord more than watchmen wait for the morning, more than watchmen wait for the morning.

Acts 17:28
For in him we live and move and have our being.

Philippians 1:20
I eagerly expect and hope that I will in no way be ashamed, but will have sufficient courage so that now as always, Christ will be exalted in my body, whether by life or death.

28 Frame of Reference

...even when we find not what we seek, we find something as well worth seeking as what we missed. - Robert Boyle, Irish chemist

A frame of reference is a perspective. We use a frame of reference to observe and describe the motion of objects. To measure and understand movement, we have to start with a frame of reference or a fixed point.

Different frames of reference can lead to different observations of the same motion. For example, if you are observing the motion of a car from a sidewalk, the sidewalk would be your frame of reference. The sidewalk would be the fixed, or non-moving, point, and you would easily see the car in motion. Whereas if you were in the car, your frame of reference would be the car, and the motion outside of the car would look different.

In the Bible, the idea of a frame of reference can be best understood by seeing things from God's point of view. We have to remember the limits of our understanding of the world around us, as well as our limited understanding of the situations we find ourselves in. Isaiah 55:8 says, "For my thoughts are not your thoughts, neither are your ways my ways," declares the Lord.

Faith should be the position we stand on. It takes faith to believe in God and move throughout our days. Hebrews 11:1 says that faith is the substance of things hoped for, the evidence of things not seen.

When we begin from a position of faith, our perspective changes, and things that need to move will move. Dare to see things from God's perspective and watch what He will do.

Psalms 93:1
The Lord reigns, he is robed in majesty; the Lord is robed in majesty and is armed with strength. The world is firmly established; it cannot be moved.

Matthew 17:20
He replied, "Because you have so little faith. I tell you the truth, if you have faith as small as a mustard seed, you can say to this mountain, 'Move from here to there' and it will move. Nothing will be impossible for you."

Matthew 19:26
Jesus looked at them and said, "With man this is impossible, but with God all things are possible."

Mark 9:23
"If you can?" said Jesus. "Everything is possible for him who believes."

Philippians 4:13
I can do everything through him who gives me strength.

29 Momentum

Destiny is no matter of chance. It is a matter of choice. It is not a thing to be waited for; it is a thing to be achieved. - William Jennings Bryan, American lawyer

Momentum can be simply defined as mass in motion. All objects have mass, so if an object is moving, then it has momentum. The amount of momentum that an object has depends on two things: how much stuff is moving and how fast the 'stuff' is moving.

A football has mass (it's made up of particles or 'stuff'). When a football is thrown, it has momentum because the mass or 'stuff' is moving. When a receiver catches a thrown football and runs with it, the speed and direction of the movement of the football change, and so does the momentum. The mass of the football didn't change, but the movement did, and that changes the overall momentum. When the football is set on the ground, there is no momentum. Objects at rest do not have any momentum because the mass is not in motion.

Momentum is an important aspect of our lives. As we travel through life, we experience momentum shifts. We slow down, experience change, and have times of rest, but we are called to keep going and not stop. In Philippians 3:13-14, Paul said, "Brothers, I do not consider myself yet to have taken hold of it. But one thing I do: forgetting what is behind and straining toward what is ahead, I press on toward the goal to win the prize for which God has called me heavenward in Christ Jesus."

And may we do the same.

Psalms 37:23-24
If the Lord delights in a man's way, he makes his steps firm; though he stumble, he will not fall, for the Lord upholds him with his hand.

Romans 12:11
Never be lacking in zeal, but keep your spiritual fervor, serving the Lord.

2 Corinthians 4:8-9
We are hard pressed on every side, but not crushed; perplexed, but not in despair; persecuted, but not abandoned; struck down, but not destroyed.

Philippians 3:12-16
Brothers, I do not consider myself yet to have taken hold of it. But one thing I do: Forgetting what is behind and straining toward what is ahead, I press on toward the goal to win the prize for which God has called me heavenward in Christ Jesus.

James 1:12
Blessed is the man who perseveres under trial, because when he has stood the test, he will receive the crown of life that God has promised to those who love him.

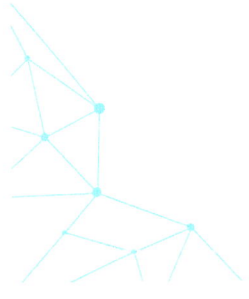

30 Force

"If A is a success in life, then A equals X plus Y plus Z - Work is X; Y is play; and Z is keeping your mouth shut." - Albert Einstein, Theoretical physicist

A force is defined as a push or pull. Gravity, friction, air resistance, and tension are a few examples of forces. A force is required to cause and change motion.

Forces always operate in pairs. The amount of force applied to an object will always equal the amount of force the object applies in return. When we walk, our feet push against the ground, and the ground pushes back in an equal but opposite direction.

When we sit in a chair, our body pushes down on the chair, but the chair pushes back with an upward force. A few force pair examples are gravity and friction, drag and air resistance, and compression and tension.

As Christians, we face opposition every day. In Deuteronomy 30:19, we can see the opposing forces that we have to choose between each day. God says, "This day I call heaven and earth as witnesses against you that I have set before you life and death, blessings and curses. Now choose life. "

We can choose to obey God's commands and follow His way, or we can resist Him, disobey, and live a life that leads to death and an eternity apart from Him.

Which will you choose today?

Psalms 34:14
Turn from evil and do good; seek peace and pursue it.

Psalms 141:4
Let not my heart be drawn to what is evil, to take part in wicked deeds with men who are evildoers, let me not eat of their delicacies.

Matthew 16:24-25
Then Jesus said to his disciples, "If anyone would come after me, he must deny himself and take up his cross and follow me. For whoever wants to save his life will lose it, but whoever loses his life for me will find it.

Romans 8:5
Those who live according to the sinful nature have their mind set on what that nature desires, but those who live in accordance with the Spirit have their minds set on what the Spirit desires.

Ephesians 6:12
For our struggle is not against flesh and blood, but against the rulers, against the authorities, against the powers of this dark world and against the spiritual forces of evil in heavenly realms.

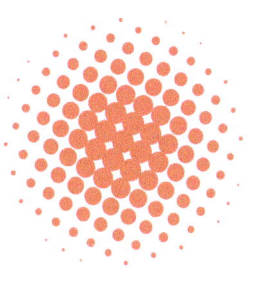

31 Fundamental Forces

Everything is theoretically impossible until it is done.
- Robert A. Heinlein, American aeronautical engineer

Science believes that there are four basic building blocks of all interactions in our universe, ranging from the smallest subatomic particles to the largest galaxies. The four building blocks are forces that are made up of gravity, electromagnetism, the weak nuclear force, and the strong nuclear force.

Gravity is the weakest of the four but acts over large distances and is a force of attraction between objects that have mass. Electromagnetism is a force that exists between electrically charged particles and includes both electrical and magnetic forces. The weak nuclear force is responsible for types of radioactive decay. It is involved in nuclear reactions. The most powerful of these four forces, the strong nuclear force, is the force responsible for holding the protons and neutrons together within an atom.

What many scientists unfortunately fail to recognize is that the ultimate force behind all four of those basic building blocks is the spoken Word of God. Hebrews 1:3 states that the Son is the radiance of God's glory and the exact representation of His being, sustaining all things by His powerful Word.

What a great revelation. From the dawn of creation, God's spoken Word created everything that exists, and He continues to hold everything together – including us. Nothing in all of the universe is more powerful than the God we serve.

Jeremiah 10:12

But God made the earth by his power; he founded the world by his wisdom and stretched out the heavens by his understanding.

Jeremiah 23:29

"Is not my word like fire," declares the Lord, "and like a hammer that breaks a rock in pieces?"

John 1:1

In the beginning was the Word, and the Word was with God, and the Word was God. He was with God in the beginning.

Matthew 4:4

Jesus answered, "It is written: 'Man does not live on bread alone, but on every word that comes from the mouth of God.'

John 17:17

Sanctify them by the truth; your Word is truth.

32 Dynamics

"Never say I tried it once and it did not work."
- Ernest Rutherford, New Zealand physicist

Dynamics is the study of why things move. It seeks to explain the causes of motion and the forces involved.

Most of us are familiar with riding a bicycle. Pedaling and balancing are the two things we probably think of when we describe how to ride a bicycle to someone else, but according to physics, riding a bicycle is much more complex. Propulsion, torque, Newton's Third Law, air and rolling resistance, steering, body movement, gravity, drag, wheels, frame, gear and chain, and many other factors are all parts of riding a bicycle.

Riding a bicycle is complicated, but we all can learn how. Being a disciple of Christ can seem like a complicated, impossible task. How could we possibly live each day meeting the expectation of a Holy God, but we can.

In the Old Testament, we read over and over about how many of God's chosen people fell short of meeting the Ten Commandments and keeping the many laws that went along with those commands.

But thank God for Jesus! He made it simple for us. Jesus said, "Love the Lord your God and love your neighbor as yourself," and He sent the Holy Spirit to live within us to give us the ability to do it.

Matthew 5:17
Do not think that I have come to abolish the Law or the Prophets; I have not come to abolish them but to fulfill them.

Matthew 5:43-45a
"You have heard that it was said, 'Love your neighbor and hate your enemy.' But I tell you: Love your enemies and pray for those who persecute you, that you may be sons of your Father in heaven.

Mark 12:30-31
Love the Lord your God with all your heart and with all your soul and with all your mind and with all your strength. The second is this: 'Love your neighbor as yourself'. There is no commandment greater than these."

John 13:34-35
"A new command I give you: Love one another. As I have loved you, so you must love one another. By this will men know that you are my disciples, if you love one another."

1 Corinthians 13:4-8a
Love is patient, love is kind. It does not envy, it does not boast, it is not proud. It is not rude, it is not self-seeking. It is not easily angered, it keeps no record of wrongs. Love does not delight in evil but rejoices with the truth. It always protects, always trusts, always hopes, always perseveres. Love never fails.

33 Contact Force

It is evident that an acquaintance with natural laws means no less than an acquaintance with the mind of God therein expressed.
  - James Prescott Joule, English physicist

Forces are divided into two categories: contact forces and field forces. To be classified as a contact force, there has to be physical touch between objects. Friction is an example of a contact force because friction is present when objects touch or rub against each other.

Field forces, on the other hand, are present whether or not objects are in contact with one another or not. These are forces like magnetism that exude a push or a pull over a particular area of space without physical touch. Field forces are much harder to understand than contact forces because we can't see or feel them.

When Jesus was crucified, the disciples were angry and confused. One particular disciple, Thomas, was not with the other disciples when Jesus visited them after He rose from the dead. When the other disciples told Thomas they had seen Jesus, he said, "Unless I see the nail marks in his hands and put my finger where the nails were, and put my hand into His side, I will not believe it."

Thomas wanted physical contact. He was only going to believe if he had physical proof. When Thomas finally saw Jesus face to face, he touched His hands and side, and he believed. Due to Thomas' lack of faith, Jesus rebuked him saying, "Because you have seen me, you have believed; blessed are those who have not seen and yet believe."

Mark 9:24
Immediately, the boy's father exclaimed, "I do believe; help me overcome my unbelief!"

John 20: 24-29
Now Thomas (called Didymus), one of the Twelve, was not with the disciples when Jesus came. So the other disciples told him, "We have seen the Lord!" But he said to them, "Unless I see the nail marks in his hands and put my finger where the nails were, and put my hand into his side, I will not believe it." A week later, his disciples were in the house again, and Thomas was with them. Though the doors were locked, Jesus came and stood among them and said, "Peace, be with you!" Then he said to Thomas, "Put your finger here; see my hands. Reach out your hand and put it into my side. Stop doubting and believe." Thomas said to him, "My Lord and my God!" Then Jesus told him, "Because you have seen me, you have believed; blessed are those who have not seen and yet have believed."

2 Corinthians 5:7
We live by faith, not by sight.

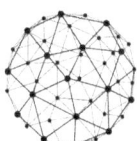

Hebrews 11:7
By faith Noah, when warned about things not yet seen, in holy fear built an ark to save his family. By his faith, he condemned the world and became heir of the righteousness that comes by faith.

Jude:20
But you, dear friends, build yourselves up in your most holy faith and pray in the Holy Spirit.

34 Inertia

The first law of motion is also known as the Law of Inertia. This law states that an object at rest will remain at rest and/or an object in motion at a constant speed will remain in motion at the same constant speed. In simplest terms, objects always resist any change in motion. This resistance to change is called inertia. If an object is resting, it does not want to move, and if an object is moving, it doesn't want to stop moving.

To overcome this resistance to change, a push or a pull must be applied. As a child of God, the last thing to change in us will be our fleshly body. In 1 Corinthians 15:51-52, Paul says, "Listen, I tell you a mystery: we will not all sleep, but we will all be changed - in a flash, in the twinkling of an eye, at the last trumpet. For the trumpet will sound, the dead will be raised imperishable, and we will be changed."

Until that day, our spirit and flesh will battle one another. We have to work hard every day to keep our flesh under submission. We must push the wrong desires away and pull the right ones towards us, and, with the help of the Holy Spirit, we can do it!

Romans 7:21-25

So I find this law at work: When I want to do good, evil is right there with me. For in my inner being I delight in God's law, but I see another law at work in the members of my body, waging war against the law of my mind and making me a prisoner of the law of sin at work within my members. What a wretched man I am! Who will rescue me from this body of death? Thanks be to God through Jesus Christ our Lord!

1 Corinthians 9:27

No, I beat my body and make it my slave so that after I have preached to others, I myself will not be disqualified for the prize.

Galatians 5:24-25

Those who belong to Christ Jesus have crucified the sinful nature, its passions, and desires. Since we live by the Spirit, let us keep in step with the Spirit.

Philippians 3:20-21

But our citizenship is in heaven. And we eagerly await a Savior from there, the Lord Jesus Christ, who, by the power that enables him to bring everything under his control, will transform our lowly bodies so that they will be like his glorious body.

1 John 3:9

No one who is born of God will continue to sin, because God's seed remains in him; he cannot go on sinning, because he has been born of God.

35 Third Law of Motion

God, the author of the universe, and the free establisher of the laws of motion.
- Robert Boyle, Irish chemist

The Third Law of Motion is also known as the Law of Action and Reaction. This law states that when two objects interact, they exert equal but opposite forces on each other. These forces have the same amount of push or pull but are acting in opposite directions.

A simple example of this law is walking. When we walk, our feet push backwards against the ground, and in response, the ground pushes forward on the feet with equal but opposite force. Rocket science also depends on this law.

When a rocket launches, the gases created by the burning fuel are forced out of the bottom of the rocket, and the opposite force is equal to the force pushing the gas out, and the rocket is thrust upward. Without these action/reaction pairs, most of the motion we experience would not happen.

There is a foundational law in the Kingdom of God that is also an action/reaction pair. It is called the Law of Sowing and Reaping. In Luke 6:38, Jesus tells, "Give and it will be given to you…For with the measure you use, it will be measured to you."

The action of sowing seeds such as kindness, forgiveness, mercy, or love into the lives of others will allow us to reap a harvest of like kind.

Hosea 8:7
They sow the wind and reap the whirlwind.

Hosea 10:12
Sow for yourselves righteousness, reap the fruit of unfailing love, and break up your unplowed ground; for it is time to seek the Lord, until he comes and showers righteousness on you.

2 Corinthians 9:6
Remember this: Whoever sows sparingly will also reap sparingly, and whoever sows generously will also reap generously.

Galatians 6:7-8
Do not be deceived: God cannot be mocked. A man reaps what he sows. The one who sows to please his sinful nature, from that nature will reap destruction; the one who sows to please the Spirit from the Spirit will reap eternal life.

Ephesians 6:7-8
Serve wholeheartedly, as if you were serving the Lord, not men, because you know that the Lord will reward everyone for whatever good he does, whether he is slave or free.

36 Tension

The meaning of life consists in the fact that it makes no sense to say that life has no meaning. - Niels Bohr, Danish physicist

Tension is a type of contact force that is transmitted through a rope, string, wire, or cable when it is being pulled or stretched. Tension force is vital to various engineering designs. When building a suspension bridge, tension is used to support the bridge's deck and the traffic that crosses it. Tension is also present in cables found in buildings that help the structure remain strong.

Fishing is another example where tension is present. Tension helps to get the most distance when casting, it keeps the line taught to "set the hook" in the fish's mouth to allow the fishermen to reel the fish in. Tension was present when Jesus chose his first disciples.

We find the account in Luke 5:1-11. Jesus was standing near the lake, and people were gathering around Him as He taught. He stepped into the boat of a fisherman named Simon and continued to teach. When Jesus finished teaching, He told Simon to go out into the deep water and cast his nets for a catch.

Simon responded by explaining that he and his crew had fished all night and caught nothing, but he would do as Jesus said. When they reached the deep and dropped their nets, they caught so many fish that the rope the net was made of was stretched to the point of breaking. The fishermen were amazed. Jesus told them, "Don't be afraid, from now on you will catch men."

This miracle took these men into their destiny. The tension of the nets held fast. God still uses tension with us today. Don't be afraid to be stretched and possibly broken - your destiny is waiting.

Psalms 119: 143
Trouble and distress have come upon me, but your commands are my delight.

Matthew 4: 18-20
As Jesus was walking beside the Sea of Galilee, he saw two brothers, Simon, called Peter, and his brother Andrew. They were casting a net into the lake, for they were fishermen. "Come follow me." Jesus said, "And I will make you fishers of men." At once, they left their nets and followed him.

Matthew 11:28-30
"Come to me, all you who are weary and burdened, and I will give you rest. Take my yoke upon you and learn from me, for I am gentle and humble in heart, and you will find rest for your souls. For my yoke is easy and my burden is light.

1 Peter 4:12-13
Dear friends, do not be surprised at the painful trial you are suffering, as though something strange were happening to you. But rejoice that you participate in the sufferings of Christ, so that you may be overjoyed when his glory is revealed.

1 Peter 5:10
And the God of all grace, who called you to his eternal glory in Christ, after you have suffered a little while, will himself restore you and make you strong, firm and steadfast.

37 Gravity

Gravity explains the motions of the planets, but it cannot explain who sets the planets in motion." - Isaac Newton, English polymath

Gravity is a field force that pulls. It is present when two objects are pulled toward each other. Gravity holds the planets in orbit around the sun. It keeps the moon in orbit around the Earth. Earth's gravity keeps us on the ground and makes things fall.

Anything that has mass has gravity; therefore, objects with more mass have more gravity. Gravity is affected by distance. Objects that are closer to each other have a stronger gravitational pull between them.

James 4:8 states, "Come near to God, and He will come near to you." As Christians, we must maintain a close relationship with God by drawing near to Him through reading His Word daily. We also draw near by singing praise and worshipping Him in Spirit and Truth. We maintain a strong relationship with Him by spending time in prayer; talking and listening to Him regularly.

The closer we walk with God, the more we get to know Him and understand His ways. Through our drawing near, He reveals Himself to us – and we begin to gain confidence in the fact that He exists, He holds us, and He will never let us go.

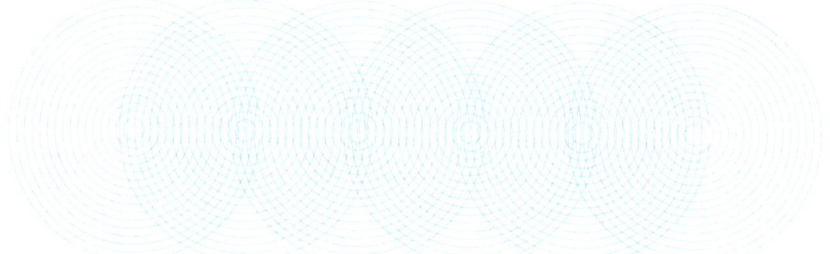

Deuteronomy 4:29
But if from there you seek the Lord your God, you will find him if you look for him with all your heart and with all your soul.

Psalms 145:18-19
The Lord is near to all who call on him, to all who call on him in truth. He fulfills the desires of those who fear him; he hears their cry and saves them.

Jeremiah 29:12-13
Then you will call upon me and come and pray to me, and I will listen to you. You will seek me and find me when you seek me with all your heart.

John 4:23-24
"Yet a time is coming and has now come when the true worshipers will worship the Father in spirit and truth, for they are the kind of worshippers the Father seeks. God is spirit, and his worshippers must worship in spirit and in truth."

Colossians 3:16
Let the word of Christ dwell in you richly as you teach and admonish one another with all wisdom, and as you sing psalms, hymns and spiritual songs with gratitude in your hearts to God.

38 Friction

Great Spirits have always encountered violent opposition from mediocre minds.
- Albert Einstein, theoretical physicist

Friction is a contact force that works against a push or a pull. There are four main types of friction: static friction, sliding friction, rolling friction, and fluid friction.

Static friction works against any force that tries to get an object to move. It is the type of force that keeps a book from sliding off a table when slightly tilted. Sliding friction occurs when an object is already in motion, like a box being pushed across the floor. Car tires rolling across the road experience rolling friction. Airplanes flying through the air or boats moving through water must overcome fluid friction to keep moving.

Jesus told us that in this world we would have trouble. This means we may encounter opposing spiritual forces with others or within ourselves that aim to slow us down or stop us. We are encouraged not to give up or quit -- to not be overcome. To overcome the friction we face, we must apply an equal and opposite force.

If God is trying to slow us down, it is our responsibility to humble ourselves (1 Peter 5:6). If the devil is trying to overcome us, James tells us to resist him and he will flee (James 4:7).

If other people are creating the friction, we are to do what Jesus taught in Matthew 4, "Love your enemies and pray for those who persecute you," and if we are wrestling within ourselves, we need to remember Psalms 46:10, "Be still, and know that I am God."

Romans 12:21
Do not be overcome by evil, but overcome evil with good.

Romans 15:5-6
May the God who gives endurance and encouragement give you a spirit of unity among yourselves as you follow Christ Jesus, so that with one heart and mouth you may glorify the God and Father of our Lord Jesus Christ.

Hebrews 10:36-39
You need to persevere so that when you have done the will of God, you will receive what he has promised. For in just a very little while, "He who is coming will come and will not delay. But my righteous one will live by faith, and if he shrinks back, I will not be pleased with him." But we are not of those who shrink back and are destroyed, but of those who believe and are saved.

1 Peter 5:6
Humble yourselves, therefore, under God's mighty hand, that he may lift you up in due time.

James 4:7
Submit yourselves, then, to God. Resist the devil, and he will flee from you.

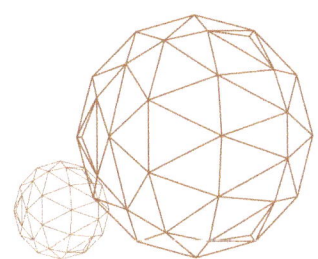

39 Power

Healing is a matter of time, but it is also a matter of opportunity.
- Hippocrates, Greek physician and philosopher

Power is the rate at which work is done or energy is transferred. In other words, power is a measure of how quickly energy is used up or passed on when work is being done. For example, if a heavy box is moved across the floor, the total amount of work is the same whether the box is pushed quickly or slowly. If the box is pushed quickly, the work is being done at a faster rate, which means more power is being used. Jesus knew we would need power to live the life He has called us to. He sent the Holy Spirit to be our power source.

In Acts 1:8, Jesus shares his last words before ascending into Heaven. He told His disciples that they would receive power when the Holy Spirit came. The word for power in the original Greek language is *dunamis*. It is translated as an inherent power, but carries a deeper meaning too: it's associated with miracles and supernatural power.

When the Holy Spirit came on the disciples, He enabled them to speak in other languages and share the gospel with boldness; 3000 people accepted the message of Christ and were baptized the very day the disciples were filled.

The same Holy Spirit that was sent to the disciples is available to us as well. When we accept the message of Christ and are baptized, God's power comes to reside in us. We can draw on His power daily to provide the energy we need to do the work He has called us to do. With His help, we will go farther and do more than we can imagine.

Isaiah 40:29
He gives strength to the weary and increases the power of the weak.

Acts 1:8
But you will receive power when the Holy Spirit comes on you, and you will be my witnesses in Jerusalem, and in all Judea and Samaria, and to the ends of the earth.

2 Corinthians 12:9
But he said to me, "My grace is sufficient for you, for my power is made perfect in weakness.

Ephesians 1:18-20
I pray also that the eyes of your heart may be enlightened in order that you may know: the hope to which he has called you, the riches of his glorious inheritance in the saints, and his incomparably great power for us to believe. That power is like the working of his mighty strength, which he exerted in Christ when he raised him from the dead and seated him at his right hand in the heavenly realms.

Ephesians 3:20-21
Now to him who is able to do immeasurably more than all we ask or imagine, according to his power that is at work within us, to him be glory in the church and in Christ Jesus throughout all generations, for ever and ever! Amen.

40 Mechanical Advantage

It is questionable if all the mechanical inventions yet made have lightened the day's toil of any human being. - John Stuart Mill, political economist

MA, short for mechanical advantage, is a measure of how much easier a machine makes work. The higher the MA, the less work we have to do. All machines provide a mechanical advantage to help us perform an array of tasks. For example, bicycle gears allow riders to pedal with differing degrees of mechanical advantage. A low gear creates more mechanical advantage and makes it easier to pedal. A high gear has less MA and is harder to pedal but assists the rider in going faster.

Mechanical advantage is a trade-off. If we reduce the effort we exert on the machine, we increase the distance the object has to travel, and vice versa. There is always a cost to the benefits of using a machine. In Luke 14:28-33, Jesus tells us there is a cost to being His disciple.

Jesus shares two stories about counting the cost of discipleship: one is about a tower being built and the other is about a king going to war. Jesus says, "Suppose one of you wants to build a tower. Will he not first sit down and estimate the cost to see if he has enough money to complete it?" and then He says, "Or suppose a king is about to go to war against another king. Will he not first sit down and consider whether he is able?" Both examples reveal the load and the effort it will take to abandon all we have to follow Him.

Everything that matters requires hard work and is costly, but the advantage of giving up everything to follow Jesus is so worth the effort.

Matthew 10:37-39
"Anyone who loves his father or mother more than me is not worthy of me; anyone who loves his son or daughter more than me is not worthy of me; and anyone who does not take his cross and follow me is not worthy of me. Whoever finds his life will lose it, and whoever loses his life for my sake will find it.

Mark 8:34-36
Then he called the crowd to him along with his disciples and said, "If anyone would come after me, he must deny himself and take up his cross and follow me. For whoever wants to save his life will lose it, but whoever loses his life for me and for the gospel will save it. What good is it for a man to gain the whole world, yet forfeit his soul?

Luke 14:25-27
"If anyone comes to me and does not hate his father or mother, his wife and children, his brothers or sisters, yes, even his own life, he cannot be my disciple. And anyone who does not carry his cross and follow me cannot be my disciple.

Hebrews 12:1-4
Therefore, since we are surrounded by such a great cloud of witnesses, let us throw off everything that hinders and the sin that so easily entangles and let us run the race with perseverance, the race marked out for us. Let us fix our eyes on Jesus, the author and perfecter of our faith, who for the joy set before him endured the cross, scorning its shame, and sat down at the right hand of the throne of God. Consider him who endured such opposition from sinful men, so that you will not grow weary and lose heart. In your struggle against sin, you have not yet resisted to the point of shedding your blood.

Revelation 21:7 He who overcomes will inherit all this, and I will be his God and he will be my son.

41 Simple Machines

The expectations of life depend upon diligence; the mechanic that would perfect his work must first sharpen his tools. - Confucius, Chinese philosopher

A simple machine is a basic tool that is used to make work easier. These simple devices change the direction of a force or change the amount of force itself.

Simple machines have been used for centuries. They are the building blocks of more complex machines and are essential for various tasks in everyday life. There are six basic simple machines: an inclined plane, a wedge, a lever, a pulley, a wheel and axle, and a screw.

An inclined plane makes moving an object from a lower place to a higher place easier like a wheelchair ramp. A wedge is made of two inclined planes placed back-to-back like an ax or scissors. A lever is a rigid bar that pivots around a fixed point. A broom, a seesaw, and belt buckles are a few examples of levers.

In the book of Ephesians, we are given a list of the tools that we have been equipped with to stand against the schemes of the devil. We have the belt of truth, the breastplate of righteousness, sandals that carry the gospel of peace, a shield to quench every fiery dart of the enemy, the helmet of salvation, and the sword of the Spirit.

These important 'simple machines' are the armor of God meant for us to put on and use every day, "so that when the day of evil comes," we will be ready.

2 Corinthians 9:8
And God is able to make all grace abound to you, so that in all things at all times, having all that you need, you will abound in every good work.

Philippians 4:19
And my God will meet all your needs according to his glorious riches in Christ Jesus.

Colossians 3:12
Therefore, as God's chosen people, holy and dearly loved, clothe yourselves with compassion, kindness, humility, gentleness, and patience.

2 Timothy 3:16-17
All Scripture is God-breathed and is useful for teaching, rebuking, correcting, and training in righteousness, so that the man of God may be thoroughly equipped for every good work.

Hebrews 13:20-21
May the God of peace, who through the blood of the eternal covenant brought back from the dead our Lord Jesus, that great Shepard of the sheep, equip you with everything good for doing his will, and may he work in us what is pleasing to him, through Jesus Christ, to whom be glory for ever and ever. Amen.

42 Wedge

The attempt is the wedge that splits its knotty way betwixt the impossible and possible.
- Alice Cary, American poet

One of the common simple machines we use in our daily lives is a wedge. This tool has the shape of a triangle from the sides because it is wide on one end and slim on the other.

Wedges push objects apart; they split and separate. Wedges can also hold objects in place. A knife, ax, and sword are wedges that separate. A door stop is a type of wedge that holds an object in place.

The Bible is full of a lot of examples that include this type of simple machine, such as sickles, axes, and swords. One example is the Word of God being compared to a sword. In Hebrews 4:12, it says, "The Word of God is living and active, sharper than any double-edged sword, it penetrates even to dividing soul and spirit, joints and marrow; it judges the thoughts and attitudes of the heart."

As representatives of Christ, our most important responsibility is to handle the sword of the Word of God correctly. But how do we do that? We begin by recognizing, like a sword, that the Word of God is not something to play around with and disrespect. We are told in 2 Timothy 3:16 that the Word corrects, rebukes, encourages, and trains us. When God reveals our wrong thoughts, attitudes, and actions through the reading and preaching of His Word, He is using His sword in one of the ways Timothy shared.

As His disciple, we must have a teachable, correctable spirit. Our loving, wise Father knows how to use the double-edged sword of His Word, and may we be humble enough to allow Him.

Job 5:17-18
"Blessed is the man whom God corrects; so do not despise the discipline of the Almighty.

Jeremiah 10:24
Correct me, Lord, but only with justice, not in your anger, lest you reduce me to nothing.

1 Corinthians 2:10-16
But God has revealed it to us by His Spirit. The Spirit searches all things, even the deep things of God. For who among men knows the thoughts of a man except the man's spirit within him? In the same way, no one knows the thoughts of God except the Spirit of God. We have not received the spirit of the world but the Spirit who is from God, that we may understand what God has freely given us.

2 Timothy 2:15
Do your best to present yourself to God as one approved, a workman who does not need to be ashamed and who correctly handles the word of truth.

2 Timothy 4:2
Preach the Word; be prepared in season and out of season; correct, rebuke, and encourage, with great patience and careful instruction.

43 Pulleys

A pulley is a simple machine that consists of a wheel with a grooved rim around it. A rope, cable, or belt passes through the groove. A pulley changes the direction of a force, making it easier to lift or move heavy objects. A pulley is very helpful when lifting objects because it allows you to pull a rope down, which requires less effort than lifting. There are a few different kinds of pulleys: a fixed pulley, a movable pulley, and a compound pulley, also known as a block and tackle. Pulleys have been used since the earliest of times.

In biblical times, water was drawn from a well using a simple pulley with a bucket attached to the rope. The empty bucket would be lowered into the water and raised full of water that was dumped into clay jars and carried to homes and animal pens. Drawing water from a well was an everyday necessity of life.

In Ecclesiastes, Solomon encourages us to remember our Creator "before the pitcher is shattered at the spring or the wheel is broken at the well." The pitcher, made of clay, weakens as it ages, and the pulley at the well will eventually break. This passage of Scripture describes our brief life: birth, growing and aging, and then eventually dying.

Like the popular saying, "Don't put off tomorrow what you can do today," your life will mean nothing if you haven't met the Creator and made Jesus the Lord of your life.

There is no better day than today to respond to the pulling in the well of your heart and draw yourself to Him now.

Psalms 95:6-8a
Come, let us bow down in worship, let us kneel before the Lord our Maker; for he is our God and we are the people of his pasture, the flock under his care. Today, if you hear his voice, do not harden your hearts.

Ecclesiastes 12:6-7
Remember him, before the silver cord is severed, or the golden bowl is broken; before the pitcher is shattered at the spring, or the wheel is broken at the well, and the dust returns to the ground it came from, and the spirit returns to God who gave it.

Isaiah 49:8-9a
This is what the Lord says: "in the time of my favor I will answer you, and in the day of salvation I will help you, I will keep you and will make you to be a covenant for the people, to restore the land and to reassign its desolate inheritances, to say to the captives, 'Come out,' and to those in darkness, 'Be free!'

Luke 12:19-21
And I'll say to myself, "You have plenty of good things laid up for many years. Take life easy; eat, drink, and be merry." "But God said to him, 'You fool! This very night, your life will be demanded from you. Then who will get what you have prepared for yourself?' "This is how it will be with anyone who stores up things for himself but is not rich toward God."

2 Corinthians 6:2
For he says, "In the time of my favor I heard you, and in the day of salvation I helped you." I tell you, now is the time of God's favor, now is the day of salvation.

44 Wheel and Axle

Nothing can be more certain than this: that we are just beginning to learn something of the wonders of the world on which we live, and move and have our being. - William Ramsay, Scottish chemist

A wheel and axle is a simple machine that consists of a rod called an axle that runs through the center of a round disk called a wheel. This simple machine transfers torque, a force that is present when something is turned, from one part to the other.

The size of the wheel determines the mechanical advantage of the wheel and axle. If the wheel has a large radius, like the Ferris Wheel, it has greater MA than a small wheel like a doorknob.

A fascinating description of a wheel is found in the book of Ezekiel. Ezekiel had a vision and tried to describe what he saw. He described seeing four living creatures with a wheel beside each creature. The wheels sparkled and appeared to be made like a wheel intersecting a wheel.

Commentary by Matthew Henry suggests the wheels are symbolic of God's divine providence. We are constantly in the protective care of God. He is omnipotent, meaning He can do anything because He is all-powerful. He is also omnipresent, meaning He can be present everywhere at the same time.

There is no aspect of our life that God is not aware of. Like a wheel that turns, we experience highs and lows, but as long as we keep Him at the center, we can be confident and secure.

We can face anything that comes because we serve the One and Only True All-Mighty, Omnipotent, All-Knowing, Omnipresent God.

Genesis 18:14
Is anything too hard for the Lord?

Psalms 33:13-14
From heaven, the Lord looks down and sees all mankind; from his dwelling place, he watches all who live on earth.

Psalms 139:1-6
O Lord, you have searched me and you know me. You know when I sit and when I rise; you perceive my thoughts from afar. You discern my going out and my lying down; you are familiar with all my ways. Before a word is on my tongue, you know it completely, O Lord. You hem me in, behind and before; you have laid your hand upon me. Such knowledge is too wonderful for me, too lofty for me to attain.

Luke 1:37
For nothing is impossible with God.

1 John 3:19-20
This then is how we know that we belong to this truth, and how we set our hearts at rest in his presence whenever our hearts condemn us. For God is greater than our hearts, and he knows everything.

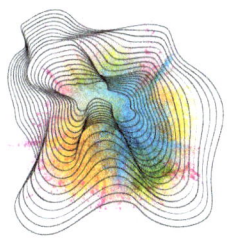

45 Thermal Energy

When you can't make them see the light, make them feel the heat.
 - Ronald Reagan, 40th US President

Thermal Energy is the energy of motion at a microscopic level. It is also known as heat energy and can be transferred between objects and systems. Thermal energy is measured by a thermometer, which tells us the temperature of whatever we are measuring. Temperature is a measurement that tells us how quickly or slowly the particles in the object are moving. As motion increases, more energy is produced in the form of heat, and the temperature will be greater.

All objects, living or nonliving, have thermal energy as long as their temperature is above absolute zero. Scientists believe that if absolute zero is ever reached, all motion would completely stop. According to NASA, cold is just the absence of heat. As long as even the slightest movement is happening, heat is present.

In Revelation 3:15, Jesus tells the church of Laodicea that He knows their deeds, that they are neither cold nor hot. He calls them lukewarm and warns that He is about to spit them out of His mouth. To be lukewarm is to be indifferent. It is equivalent to having a half-hearted devotion to God; being neither fully committed nor fully opposed to Him.

This type of spiritual apathy is displeasing to God. Jesus issues this warning to a church headed towards missing the mark. I pray that we would take our spiritual temperature and do what is necessary to stay fully committed to the Lord.

Luke 6:46
"Why do you call me, Lord, Lord, and do not do what I say?"

2 Timothy 4:3-4
For the time will come when men will not put up with sound doctrine. Instead, to suit their own desires, they will gather around them a great number of teachers to say what their itching ears want to hear. They will turn their ears away from truth and turn aside to myths.

Titus 1:15-16
To the pure, all things are pure, but to those who are corrupted and do not believe, nothing is pure. In fact, both their minds and consciences are corrupted. They claim to know God, but by their actions they deny him. They are detestable, disobedient, and unfit for doing anything good.

1 John 2:15-16
Do not love the world or anything in the world. If anyone loves the world, the love of the Father is not in him.

Revelation 3:15-16
I know your deeds, that you are neither cold nor hot. I wish you were either one or the other! So, because you are lukewarm, neither hot nor cold, I am about to spit you out of my mouth.

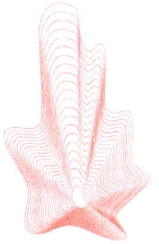

46 Fluid

The bigger the dam of patience, the worse the flood when the dam breaks.
- Austin O'Malley, author

A fluid is a substance that can flow, meaning fluids can move and change shape easily. Fluids don't have a fixed shape and readily change shape when a force is applied. Liquids, gases, and plasmas are all fluids.

There are several physical properties of fluids: viscosity, density, surface tension, compressibility, buoyancy, and pressure. These properties influence how fluids interact with their environment and guide their behavior.

For example, all fluids exert pressure. Fluid pressure is exerted equally in all directions at any given point. Pressure in a fluid increases with depth. When engineers build a dam, the base of the dam is extremely thick to withstand the extreme pressure at the bottom.

Proverbs 17:14 says, 'Starting a quarrel is like breaching a dam; so drop a matter before a dispute breaks out.' Fighting, arguing, or stirring up strife is like releasing a flood. It is similar to a dam that has been breached due to a buildup of pressure that can't be contained.

As Christians, we must learn how to overcome the selfish desires that well up from within. It takes humility and deciding to stop a quarrel before we create the cracks that can cause the dam to break. Thankfully, even if we think it's an impossible task, we can do it because we have the help of the Holy Spirit living inside us.

Proverbs 15:18

A hot-tempered man stirs up dissension, but a patient man calms a quarrel.

1 Corinthians 3:3

You are still worldly. For since there is jealousy and quarreling among you, are you not worldly? Are you not acting like mere men?

2 Timothy 2:23-24

Don't have anything to do with foolish and stupid arguments, because you know they produce quarrels. And the Lord's servant must not quarrel; instead, he must be kind to everyone, able to teach, and not resentful.

Titus 3:9-10

But avoid foolish controversies and genealogies and arguments and quarrels about the law, because these are unprofitable and useless.

James 4:1-3

What causes fights and quarrels among you? Don't they come from your desires that battle within you? You want something but don't get it. You kill and covet, but you cannot have what you want. You quarrel and fight. You do not have, because you ask with wrong motives, that you may spend what you get on your pleasures.

47 Surface Tension

Miracles are a retelling in small letters of the very same story which is written across the whole world in letters too large for some of us to see. - C.S. Lewis, British writer

Liquids experience a unique property called surface tension. Surface tension is created by the molecules that make up the liquids. These molecules attract the other molecules below and beside them, creating an inward pull. This inward pull contracts the surface area, causing the liquid to behave like a stretched elastic membrane.

The strength of surface tension depends on the particular type of liquid. Water has high surface tension, which is why water forms droplets and why some insects can walk across water.

One of the miracles Jesus performed was to walk on water. He and Peter are the only two men to have ever been a part of such a miracle. We find the account in Matthew, Mark, and John. This miracle occurred shortly after Jesus multiplied five loaves of bread and two fish to feed over 5,000 people. When everyone had eaten until they were full, Jesus sent the disciples ahead of Him. He dismissed the crowd and spent some time alone in prayer. When evening came, He observed the disciples struggling with their oars because the wind was against them. He then went out to them, walking on the water. When they saw Him, they were scared because they thought He was a ghost. He identified Himself and told them not to be afraid.

From this miraculous event, we can learn some important truths: First, God doesn't always show up the way we think He will. Secondly, we must learn to recognize Him in our storms, and lastly, He will get us to the other side if we just believe and stay focused on Him.

Matthew 8:23-27

Then he got into the boat, and his disciples followed him. Without warning, a furious storm came up on the lake, so that the waves swept over the boat. But Jesus was sleeping. The disciples went and woke him, saying, "Lord, save us! We're going to drown!" He replied, "You of little faith, why are you so afraid?" Then he got up and rebuked the winds and the waves, and it was completely calm. The men were amazed and asked, "What kind of man is this? Even the winds and the waves obey him!"

Matthew 14:25-32

During the fourth watch of the night, Jesus went out to them, walking on the lake. When the disciples saw him walking on the lake, they were terrified. "It's a ghost," they said, and cried out in fear. But Jesus immediately said to them: "Take courage! It is I. Don't be afraid." "Lord, if it's you," Peter replied, "tell me to come to you on the water." "Come," he said. Then Peter got down out of the boat, walked on the water, and came toward Jesus. But when he saw the wind, he was afraid and, beginning to sink, cried out, "Lord, save me!" Immediately, Jesus reached out his hand and caught him. "You of little faith," he said, "why did you doubt?"

Mark 4:40

He said to his disciples, "Why are you so afraid? Do you still have no faith?"

Mark 6: 45-52

Immediately, he spoke to them and said, "Take courage! It is I. Don't be afraid."

John 6:20-21

But he said to them, "It is I; don't be afraid."Then they were willing to take him into the boat, and immediately the boat reached the shore where they were heading.

48 Magnetism

But the law of magnetism really is true; who you are is who you attract.
- John C. Maxwell, American author

Magnetism is a field force. It is produced by moving electric charges. Magnets are what we call the metal or rock that contains these electric charges. Magnets can attract or repel other magnets without touching them. They can also change the motion of other charged particles. Magnets are only attracted to the metals iron, cobalt, and nickel, and any object that contains one or more of these metals.

All magnets have two poles: a north pole and a south pole. These poles are the regions where the force is the strongest. Opposite poles attract each other, and like poles repel each other. Magnets can also magnetize other metals by causing the electrons in the metal to align, creating a temporary magnetic field.

Similar to the attractive property of magnets, Jesus says, "No one can come to me unless the Father Who sent Me draws Him." We find in another recording of Jesus, He says that when He is lifted up from the Earth, He will draw all men to Himself.

No one understood what He was talking about at the time, but we know it today. He chose to die for mankind, and in doing so, His loving sacrifice attracts us to Him. As we stay close to Him, He will rub off on us, and then others that He brings into our lives will be attracted to Him too.

Jeremiah 31:3
The Lord appeared to us in the past, saying: "I have loved you with an everlasting love; I have drawn you with loving-kindness."

John 6:44
"No one can come to me unless the Father who sent me draws him, and I will raise him up at the last day."

John 12:32
"But I, when I am lifted up from the earth, will draw all men to myself."

Hebrews 10:23-24
Let us hold unswervingly to the hope we profess, for he who promised is faithful. And let us consider how we may spur one another on toward love and good deeds.

James 4:8
Come near to God and he will come near you.

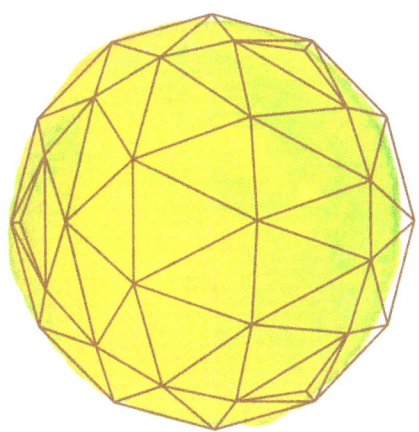

49 Electric Circuits

It's not enough to wire the world if you short-circuit the soul.
 - Tom Brokaw, American author

An electric circuit is made of a conductive material, usually a metal wire, that is connected in a continuous loop that allows electrons to flow through the material.

There are direct current circuits and alternating current circuits. In a direct current, one end of the circuit is positive and the other end is negative. In an alternating current circuit, the polarity of each pole alternates between positive and negative. In either circuit, electrons flow through the wire from the negative pole to the positive pole.

For a circuit to work properly, there has to be one or more of the following parts: a battery or energy source and a wire. These items would then be connected to whatever needs power, like a light bulb or a motor.

God has given us a type of circuit called prayer. Prayer allows us to connect with God, our Source of supply. We are told to pray in the Spirit on all occasions with all kinds of prayers and requests.

There are nine types of prayers mentioned in the Bible: Prayers of worship, thanksgiving, petition, intercession, agreement, faith, dedication, judgment, and praying in the Spirit. Prayer is a vital part of the Christian life. We must ensure the prayer line remains connected, allowing Him to work in and through us.

Psalms 95:2-3
Let us come before him with thanksgiving and extol him with music and song. For the Lord is the great God, the great King above all gods.

1 Corinthians 14:14-15
For if I pray in a tongue, my spirit prays, but my mind is unfruitful. So what shall I do? I will pray with my mind; I will sing with my spirit, but I will also sing with my mind.

Philippians 4:6
Do not be anxious about anything, but in everything, by prayer and petition, with thanksgiving, present your requests to God.

1 Timothy 2:1
I urge, then, first of all, that requests, prayers, intercession, and thanksgiving be made for everyone.

James 5:15
And the prayer offered in faith will make the sick person well; the Lord will raise him up. If he has sinned, he will be forgiven.

50 Grounding

Electricity can be dangerous. My nephew tried to stick a penny into a plug. Whoever said a penny doesn't go far didn't see him shoot across the floor. I told him he was grounded. - Tim Allen, American actor

Grounding is a safety measure used in electrical systems. It helps prevent electric shock, fires, and damage to equipment. Grounding provides a safe path for electricity to flow.

Most circuits are grounded by taking a conductive wire attached to a grounding rod and driving it into the Earth. When electricity is grounded, it returns to its source if it encounters a problem in the circuit.

In Paul's letter to the Ephesians, he prays that we, "being rooted and grounded in love," may have the power to grasp the love of Christ. Think of being grounded like a tree being planted. If we allow our roots to grow deep, we become settled, stable, and strong, not given to the trials and tribulations that can easily uproot our faith.

Like a grounded wire, being grounded in love causes us to return to The Source when we get overloaded or overwhelmed. It is from The Source where we find exactly what we need to begin again.

Isaiah 41:10
So do not fear, for I am with you; do not be dismayed, for I am your God. I will strengthen you and help you; I will uphold you with my righteous right hand.

John 15:9
As the Father has loved me, so have I loved you. Now remain in my love.

Ephesians 3:17-19
So that Christ may dwell in your hearts through faith. And I pray that you, being rooted and established in love, may have power, together with all the saints, to grasp how wide and long and high and deep is the love of Christ, and to know this love that surpasses knowledge, that you may be filled to the measure of all the faithfulness of God.

Colossians 2:6-7
So then, just as you received Christ Jesus as Lord, continue to live in him, rooted and built up in him, strengthened in the faith as you were taught, and overflowing with thankfulness.

James 1:21
Therefore, get rid of all moral filth and the evil that is so prevalent and humbly accept the word planted in you, which can save you.

51 Sound

"It would be possible to describe everything scientifically, but it would make no sense; it would be without meaning, as if you described a Beethoven Symphony as variations of wave pressure" - Albert Einstein, theoretical physicist

Sound is a form of energy. It is the only type of energy that requires a medium, like air or a solid material, to travel through. Sound is created by a series of vibrations that disturb the particles that surround an object. For example, when you pluck the string on a guitar, the surface of the string vibrates; it pushes and pulls on the air molecules around the string. The vibrations cause the molecules of air to get closer together (compression) and spread out (rarefaction). This alternating compression and rarefaction creates the traveling sound waves.

Sound waves can bounce off surfaces and bend around obstacles. They can change direction as they pass through different materials (also called mediums) and can be absorbed, which reduces their intensity.

The psalmist tells us to sing joyfully to the Lord; to praise Him with the harp and the lyre, an instrument similar to a guitar. Praise is a sound we create, an offering of thanksgiving, or a sacrifice we offer when we don't feel like singing. We were created to praise.

Praise gets things moving. Paul and Silas sang songs of praise in prison, and the chains were loosed. The walls of Jericho fell when the people raised their voices. Praise silences the enemy and can change the atmosphere of the place we are in.

Praise the Lord at all times and watch what He will do.

Joshua 6:20
When the trumpets sounded, the people shouted, and at the sound of the trumpet, when the people gave a loud shout, the wall collapsed; so every man charged straight in, and they took the city.

Psalms 8:2
From the lips of children and infants, you have ordained praise because of your enemies, to silence the foe and the avenger.

Psalms 33:1-3
Sing joyfully to the Lord, you righteous; it is fitting for the upright to praise him.

Psalms 150:6
Let everything that has breath praise the Lord.

Acts 16:25
About midnight, Paul and Silas were praying and singing hymns to God, and the other prisoners were listening to them.

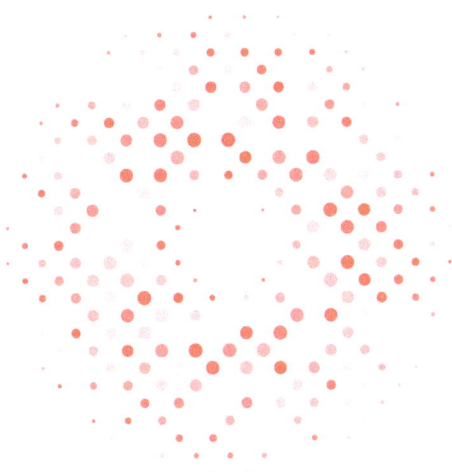

52 Light

Light is not so much something that reveals as it is itself the revelation.
- James Turrell, American artist

Light is dual. It is both a wave and a particle. The formal scientific name for light is electromagnetic radiation. Light travels at a constant speed of 186,000 miles per second.

Light has many different forms. There is the light we can see represented by the colors of the rainbow. Red has long wavelengths, and violet has short wavelengths. The colors between red and violet each have their wavelength size as well.

There is also light that the human eye cannot see. Wavelengths longer than red are called infrared, microwaves, and radio waves. Wavelengths shorter than the color violet are called ultraviolet and gamma rays.

All of these traveling wavelengths are made of particles of energy called photons. The higher the wavelength, the more energy it carries. Gamma waves have a lot more energy than radio waves. The word photon comes from the ancient Greek word *phos* or *photos*, which means light.

Phos is a general term that includes both natural light and divine light. In Revelation, we read about the New Jerusalem in chapters 21 and 22. God tells us there will be a day when the natural light will be replaced or overshadowed by the Divine Light because God's presence will be revealed. We will see Him face to face in all his glory, and I can't wait…how about you?

John 1:4-5
In him was life, and that life was the light of men. The light shines in the darkness, but the darkness has not understood it.

1 Timothy 6:14-16
I charge you to keep this command without spot or blame until the appearing of our Lord Jesus Christ, which God will bring about his own time, God, the blessed and only Ruler, the King of kings and Lord of lords, who alone is immortal and who lives in unapproachable light, whom no one has seen or can see. To him be honor and might forever. Amen.

1 John 1:5-7
This is the message we have heard from him and declare to you: God is light; in him there is no darkness at all. If we claim we have fellowship with him yet walk in the darkness, we lie and do not live by the truth. But if we walk in the light, as he is in the light, we have fellowship with one another, and the blood of Jesus, his Son, purifies us from all sin.

Revelations 21:22-23
I did not see a temple in the city, because the Lord God Almighty and the Lamb are its temple. The city does not need the sun or the moon to shine on it, for the glory of God gives it light, and the Lamb is its lamp.

Revelations 22:3-5
There will be no more night. They will not need the light of a lamp or the light of the sun, for the Lord God will give them light. And they will reign for ever and ever.

———————————————————

Prayer of Salvation

Lord Jesus,

I believe you are the Son of God, born of the Virgin Mary. I believe you shed your blood and died for me. I believe you were dead and buried and rose to life again. I confess that I am a sinner and I repent of my sins. From this day forward, I choose to walk in your ways. Thank you for coming into my life and saving me. I declare that as of today, you are the Lord of my life and my Savior forevermore.

If you prayed this prayer, take the next step and let someone know, and then find a local church and begin the greatest adventure of your life.